VECTOR BUNDLES IN MATHEMATICAL PHYSICS
VOLUME 1

MATHEMATICS LECTURE NOTE SERIES

J. Frank Adams	LECTURES ON LIE GROUPS
E. Artin and J. Tate	CLASS FIELD THEORY
Michael Atiyah	K-THEORY
Jacob Barshay	TOPICS IN RING THEORY
Hyman Bass	ALGEBRAIC K-THEORY
Melvyn S. Berger Marion S. Berger	PERSPECTIVES IN NONLINEARITY
Armand Borel	LINEAR ALGEBRA GROUPS
Raoul Bott	LECTURES ON K (X)
Andrew Browder	INTRODUCTION TO FUNCTION ALGEBRAS
Gustave Choquet	LECTURES ON ANALYSIS I. INTEGRATION AND TOPOLOGICAL VECTOR SPACES II. REPRESENTATION THEORY III. INFINITE DIMENSIONAL MEASURES AND PROBLEM SOLUTIONS
Paul J. Cohen	SET THEORY AND THE CONTINUUM HYPOTHESIS
Eldon Dyer	COHOMOLOGY THEORIES
Robert Ellis	LECTURES ON TOPOLOGICAL DYNAMICS
Walter Feit	CHARACTERS OF FINITE GROUPS
John Fogarty	INVARIANT THEORY
William Fulton	ALGEBRAIC CURVES
Marvin J. Greenberg	LECTURES ON ALGEBRAIC TOPOLOGY LECTURES ON FORMS IN MANY VARIABLES
Robin Hartshorne	FOUNDATIONS OF PROJECTIVE GEOMETRY
Robert Hermann	FOURIER ANALYSIS ON GROUPS AND PARTIAL WAVE ANALYSIS LECTURES IN MATHEMATICAL PHYSICS, VOL. I LIE ALGEBRAS AND QUANTUM MECHANICS
J. F. P. Hudson	PIECEWISE LINEAR TOPOLOGY
Irving Kaplansky	RINGS OF OPERATORS
K. Kapp and H. Schneider	COMPLETELY O-SIMPLE SEMIGROUPS

Joseph B. Keller Stuart Antman	BIFURCATION THEORY AND NONLINEAR EIGENVALUE PROBLEMS
Serge Lang	ALGEBRAIC FUNCTIONS RAPPORT SUR LA COHOMOLOGIE DES GROUPS
Ottmar Loos	SYMMETRIC SPACES I. GENERAL THEORY II. COMPACT SPACES AND CLASSIFICATIONS
I. G. Macdonald	ALGEBRAIC GEOMETRY: INTRODUCTION TO SCHEMES
George W. Mackey	INDUCED REPRESENTATIONS OF GROUPS AND QUANTUM MECHANICS
H. Matsumura	COMMUTATIVE ALGEBRA
Andrew Ogg	MODULAR FORMS AND DIRICHLET SERIES
Richard Palais	FOUNDATIONS OF GLOBAL NON-LINEAR ANALYSIS
William Parry	ENTROPY AND GENERATORS IN ERGODIC THEORY
D. S. Passman	PERMUTATION GROUPS
Walter Rudin	FUNCTION THEORY IN POLYDISCS
David Russell	OPTIMIZATION THEORY
Jean-Pierre Serre	ABELIAN *l*-ADIC REPRESENTATIONS AND ELLIPTIC CURVES ALGEBRES DE LIE SEMI-SIMPLE COMPLEXES LIE ALGEBRAS AND LIE GROUPS
Shlomo Sternberg	CELESTIAL MECHANICS PART I CELESTIAL MECHANICS PART II
Moss E. Sweedler	HOPF ALGEBRAS

A Note from the Publisher

This volume was printed directly from a typescript prepared by the author, who takes full responsibility for its content and appearance. The Publisher has not performed his usual functions of reviewing, editing, typesetting, and proofreading the material prior to publication.

The Publisher fully endorses this informal and quick method of publishing lecture notes at a moderate price, and he wishes to thank the author for preparing the material for publication.

VECTOR BUNDLES IN MATHEMATICAL PHYSICS
VOLUME I

ROBERT HERMANN

W. A. BENJAMIN, INC.

New York 1970

VECTOR BUNDLES IN MATHEMATICAL PHYSICS VOLUME I

Standard Book Numbers: 8053-3946-9 (C)
8053-3945-0 (P)

Library of Congress Catalog Card Number 72-135363
Manufactured in the United States of America
12345 R 43210

W. A. BENJAMIN, INC.
New York, New York 10016

Preface

As in my earlier book "Lie groups for physicists" [1], I mean these notes to introduce to physicists, other scientists, and perhaps even to mathematicians certain ideas arising out of contemporary Lie group theory and differential geometry that may be helpful to them in their own work and may contribute towards the description of the physical world in mathematical terms. (Perhaps God is a Lie group or a vector bundle?)

The idea of a "vector bundle"—usually also associated with the differentiable manifold idea—is one of the keystones of contemporary mathematics, appearing in the theory of partial differential equations (see Palais [1, 2]), algebraic topology (see Atiyah [1] and Husemoller [1]), and algebraic and differential geometry. It expresses some of the most basic intuitive geometric ideas in precise and conceptually clear terms, and the mathematical formalism developed over the years for the study of vector-bundle related concepts often leads most directly and simply to the clarification and solution of those problems that can be phrased in vector-bundle language.

As was pointed out already in "Lie groups for physicists," the vector bundle idea is also very natural in physics, appearing—at least implicitly—throughout both the "classical" and "quantum" theory fields, for example. Indeed, the very notion of a "field" itself—as much as it can be formalized—clearly involves "vector bundle" ideas, since one thinks of a "field" in primitive terms as a function assigning to each point of "our" space several numbers, the "field parameters" at that point. In terms of our jargon, this would translate into a definition of a "field" as a cross-section of a vector bundle, whose "base" is "our" space, and whose fibers are vectors describing the possible values of "fields" at points of space.

As in "Lie groups for physicists," I hope that a presentation that is more informal than is customary in mathematics books and that is interspersed with physical applications and ideas will help the reader penetrate more easily into the interesting concepts provided by contemporary mathematics. In this approach, "proofs" are not important for their own sake, but only as guideposts into new territory. In general, I think it is useful to keep in mind Poincaré's division of mathematicians into two categories, the "geometers" and "analysts." Now, this does *not* mean people who study geometry and analysis; but refers instead to a division between those who prefer one type of reasoning to another; say, those who like broad, qualitative, intuitive, i.e. geometry-like arguments, and those who treasure precise completely logical, deductive sorts of mathematics. Presumably, this arose in Poincaré's mind to explain his disagreements in matters of taste between his contemporaries like Cantor and Weirstrass, who were engaged in the process of driving out from the foundations of mathematics all the sloppiness that Poincaré found so stimulating.

Today, the "analysts" hold power. Most mathematicians recognize and practice only work that produces hard-headed, precise theorems that are "deep" and technically difficult, and that contribute to the solution of a problem that is recognized—by a consensus of mathematicians—as important and difficult. (If it is a "famous," old problem so much the better.) It is in the nature of things that a "geometer" will not do this, at least as his main activity. It seems to me, however, that if mathematics is ever to interact again with the outside world—say, with theoretical physics, at least—we must revive the "geometry" spirit, working in, of course, the advances in technique and formalism that have been provided by the labors of the "analysts." This book is written in that spirit.

When writing "Lie groups for physicists" I could assume that the physicist reader already had some familiarity with groups and their representations, since some of this material is taught in quantum mechanics, elementary particle, and solid-state courses, and there existed several excellent review articles by physicists which bridged the gap between my point of view and that held by most physicists. In this book, I must confess to a feeling of less confidence in my position, since it will be necessary to use, at some point in the development, the formalism of differentiable manifold theory. Now, I am confident that—in the long run—this will be taught to every student interested in physical applications, and even at the undergraduate-advanced calculus level. (This trend is exemplified by—and no doubt will be stimulated by—the recent books by Spivak [1] and Loomis and Sternberg [1].) At the moment, all I can suggest is that the reader who is completely ignorant of differentiable manifold theory read at least poetically the introductory material that does exist (Loomis and Sternberg [1], Auslander and McKenzie [1], Flanders [1]), possibly consult the review articles written by physicists involving differentiable manifolds—mostly in the literature of General

Relativity theory of the last five years (for example, see Trautman [1])—and consult the introductory parts of my book on differential geometry [2]. If he perseveres, the reader will ultimately find that the formalism provides a marvelously unified set of tools for the description of those natural phenomena that involve a combination of geometry and the differential–integral calculus. Finally, the reader might find some of the material in my monographs [3, 4] and series of articles [5] useful to supplement what is provided in this book.

I am indebted to M. Kac and H. P. McKean for their hospitality at Rockefeller University, while this book was written. I would also like to thank Mrs. Alta Zapf, for her excellent typing.

New York, New York
1970

Robert Hermann

Contents

CHAPTER I

GENERAL CONCEPTS AND NOTATIONS

We shall begin by presenting some of the general ideas in a relatively abstract form; talking of general things like "spaces", "mappings", "fibers", etc. Of course, the situations of most interest in physics are much more special, but it is really very little extra work to sketch how things go in general.

The proper framework for the discussion we are about to give is that of "topological spaces and continuous maps". Mathematicians are exposed to this in the course of their training. Since physicist readers may not be, we will not be explicit about such topological assumptions - except where it is necessary for technical purposes.

1. SPACES AND MAPPINGS

We will be dealing with a certain class of spaces. M will typically denote such a space, and p will denote a typical point of M.

A *mapping* between two such spaces M' and M is a function whose domain is M', and whose values lie in M. If ϕ denotes the mapping, we often write "ϕ: $M' \to M$ " to indicate exactly what is what. If $p \in M$, the inverse image set:

$$\phi^{-1}(p) = \text{set of } p' \in M' \text{ such that } \phi(p') = p$$

is called the *fiber of* ϕ *above* p.

Two broad classes of such mappings occur most often in geometric or physical situations, *injections* and *projections*. ϕ is called an *injection* if ϕ is one-one, i.e. if $\phi(p_o') \neq \phi(p')$ for any two distinct points p_o', p' of M. ϕ is called a *projection* if ϕ is onto M, i.e. if $\phi(M') = M$, where $\phi(M')$ is the image set of M'.

Most spaces of interest are subsets of Euclidean spaces. It is then appropriate to indicate the notations we will use for them.

R denotes the real numbers. R^n denotes the Cartesian product of n copies of R. In other words, an element of R^n - denoted, say, by x - is an ordered n-tuple $(x_1, \ldots, x_n)$ of real numbers. A typical example of a "projection" map would be to send $R^n \to R^{n-1}$ as follows:

$$\phi(x_1, \ldots, x_n) = (x_1, \ldots, x_{n-1}).$$

Similarly, a typical "injection" $R^{n-1} \to R^n$ might be:

$$\phi(x_1, \ldots, x_{n-1}) = (0, x_1, \ldots, x_{n-1}).$$

There is an abstract pattern to these constructions which it is useful to keep in mind. Start off with two spaces, denoted, say, by M' and M". Let M be the Cartesian product M' × M". Recall that a point p of M is an ordered pair (p', p") of points; p' ε M', p" ε M". One can then project M onto M' or M" by assigning

$$\phi(p', p'') = p' \quad \text{or} \quad p''.$$

Given $p_o' \in M'$, one can inject M'' into M by assigning (p_o', p'') to $p'' \in M''$.

It is often useful to construct the "graph" of a mapping ϕ': $M' \to M''$. To do so, set $M = M' \times M''$. Let ϕ be the mapping: $M' \to M$ defined by:

$$\phi(p') = (p', \phi'(p'))$$

ϕ is called the *graph* of ϕ'. Notice that ϕ is injective.

Of course, if ϕ is an "injective" map of M' into M, it is often useful to weaken the precision, and identify M' with its image $\phi(M')$ in M.

2. FIBER MAPPINGS AND VECTOR BUNDLES

Change notations slightly: E and M are spaces, with π: $E \to M$ a projective mapping of E onto M. Consider also another pair (E', M') of such spaces, with π': $E' \to M'$ a projective mapping.

<u>DEFINITION</u>. A map ϕ: $E \to E'$ is *fiber preserving*

if, for each $p \in M$, ϕ maps the fiber $\pi^{-1}(p)$ into one of the fibers of the map π'.

Consider such a fiber preserving map ϕ. For $p \in M$, suppose that $\phi(\pi^{-1}(p))$ is contained in $\pi'^{-1}(p')$, for some point $p' \in M'$. Notice that p' is uniquely determined by p, so that we can define a map ϕ_M: $M \to M'$ by assigning $p' = \phi_M(p)$ to $p \in M$. In a formula,

$$\phi_M(p) = \pi'(\phi(\pi^{-1}(p)))$$

$$\text{for} \quad p \in M.$$

This map ϕ_M is said to be the *quotient* map of ϕ relative to the projections π, π'.

<u>DEFINITION</u>. The projective mappings (π, E, M), (π', E', M') are said to be *isomorphic* relative to the fiber preserving map ϕ if ϕ and ϕ_M are one-one, onto maps, i.e. if inverse maps ϕ^{-1}, $\phi_M{}^{-1}$ exist. (Of course, if a topological, algebraic, or differential structure is imposed, one would also require - as implied by using the word "isomorphism" - that these maps preserve these additional structures. For example, if "topology" is imposed, one would

require that ϕ, ϕ_M be homeomorphisms.)

DEFINITION. A projective map π: $E \to M$ is *isomorphic to the product* (or is said to "have a product structure") if fiber isomorphism ϕ: $E \to E'$ can be chosen, for the special choice:

$E' = M' \times M$, the Cartesian product

π': $E' \to M$ is the Cartesian projection.

Another way of putting this is to say that there is a map α: $E \to M'$ such that:

For each point $p \in M$, α is a one-one, onto map of the fiber $\pi^{-1}(p)$ onto M'. (2.1)

With such an α, the fiber space isomorphism ϕ: $E \to M \times M'$ is defined by:

$$\phi(x) = (\pi(x), \alpha(x)) \quad (2.2)$$

for $x \in E$.

This description of a product fiber space is most convenient for the purposes of differential geometry, where a subject of prime importance is the question of independence of various concepts on the *choice* of the product-isomorphisms. Suppose then that α, α' are two maps: $E \to M'$ satisfying (2.1). Then, because of (2.1), for each point $p \in M$ there is a transformation[1] $\beta(p)$: $M' \to M'$ such that:

$$\alpha(x) = \beta(p)(\alpha'(x)) \tag{2.3}$$

for each point $x \in \pi^{-1}(p)$. Then, the "transition maps" $\beta(p)$ provide a map $p \to \beta(p)$ of M into the group of all transformations of M' onto itself.

Now, we can define the key concept of "fiber space". (For the sake of simplicity, we will restrict attention to what are, technically, called "local product fiber spaces". More general sorts are of great interest in a wide variety of problems in differential geometry and topology, but most of the obvious examples of "fiber spaces" are, indeed,

---

[1]By a *transformation* of a space M' we will mean a one-one map of M' onto itself. Thus, the inverse map exists. The set of all transformations form a group, with the group operation just composition of maps. If M' is a topological space, we will usually also want to restrict attention to such transformations β: $M' \to M$ that are homeomorphisms, i.e. the maps β, β^{-1} are continuous.

"local product".) Suppose that π: $E \to M$ is a projective map. It will be convenient now to use the terminology of point - set topology: Suppose that E and M are topological spaces, and π is a continuous map. Suppose that M' is another topological space, fixed in advance. Suppose that U is an open subset of M.

$$E_U = \pi^{-1}(U),$$

i.e. E_U is the set of points of E that "project" under π into points of U. Then, (E_U, U, π) is a projective map. (E, M, π) is said to be a *product over U* if there is a product isomorphism ϕ_U: $E_U \to U \times M'$ which is also a homeomorphism in the point - set topology sense. Let α_U: $E_U \to M'$ be the map defined by (2.2), and the discussion preceding (2.2).

<u>DEFINITION</u>. (E, M, π) is a *fiber space*, with fiber M', if M has a covering by open subsets, denoted typically by U, U', U",..., such that (E, M, π) is a product over U, as defined above.
(Recall that a collection of subsets of M is a

covering of each point of M is in at least one point of M. Of course, individual points of M may lie in any number of subsets. Usually, we will work with situations where each point lies in only a finite number of the subsets).

Let us make explicit the resulting ambiguity of description of points of E as elements of M × M'. Suppose that U, U' are two subsets of M, that intersect in a subset $U \cap U'$. Recall that $U \cap U'$ is also an open subset of M. Then, α_U: $E_U \to M$, $\alpha_{U'}$: $E_{U'} \to M'$ restrict to $E_{U \cap U'}$, resulting in two maps into M. There is then - via (2.2) - a map:

$$\beta_{U,U'}: \ U \cap U' \to \text{(group of homeomorphisms of } M')$$

such that:

$$\alpha(x) = \beta_{U,U'}(\pi(x))(\alpha'(x)) \qquad (2.4)$$

for all points $x \in E_{U \cap U'} = \pi^{-1}(U \cap U')$.

The defining relations (2.4) show that, in the overlap $U \cap U' \cap U''$ of three open sets of the covering, the system of "transition maps" $\{\beta_{U, U'}\}$ satisfy a system of functional equations, related to what the topologists call a "cocycle condition" namely:

$$\beta_{U,U'} = \beta_{U,U''}\, \beta_{U'',U} \qquad (2.5)$$

Conversely, such a system of maps $\{\beta_{U,U'}\}$ arises from a fiber space. This approach was used by Steenrod [1] in his pioneering exposition of the theory of fiber spaces, but the formalism resulting from an emphasis on this point of view is unduly complicated for the applications we have in mind; although it will be occasionally useful. The most useful fact in this approach is the constructive definition of E in terms of the transition maps, which we will briefly describe.

<u>DEFINITION</u>. Given two spaces A, A' of points, the *disjoint union* of A and A' is the space whose points are those of A *or* A', with points of A and A' considered as completely different from each other. For example, if points of A and A' are countable, labelled as $(a_1, a_2, \ldots)$ and

$(a_1', a_2', \ldots)$, the points of the disjoint union are labelled as $(a_1, a_1', a_2, a_2', \ldots)$.

Suppose now that $\{U, U_0', U', \ldots\}$ is an open covering of M, M' is a space, and $\{\beta_{U,U'}\}$ is a system of maps: $U \cap U' \to$ (group of homeomorphisms of M'), satisfying (2.5) with overlap of any three subsets. Let:

$$A = \text{disjoint union of } U \times M,\ U' \times M', \ldots .$$

Now, introduce an equivalence relation[1] in A as follows:

---

[1]Recall this definition: If A is a space, an *equivalence relation* in A is a subset B of A × A such that: a) If $(a_1, a_2) \in B$ for $a_1, a_2 \in A$, then $(a_2, a_1) \in B$, b) If (a_1, a_2) and $(a_2, a_3) \in B$, then $(a_1, a_3) \in B$. It is often customary to write $a_1 \sim a_2$ (real "a_1 equivalent to a_2") if $(a_1, a_2) \in B$.

$$(p, q) \in U \times M \text{ is equivalent to } (p', q') \in U' \times M' \text{ if and only if:} \qquad (2.6)$$

a) $p = p'$

b) $q = \beta_{U\,U'}(p)(q')$

One verifies that the condition (2.5) is precisely that needed to show that (2.6) is a genuine equivalence relation. The quotient[1] of A by this equivalence relation is defined as E. Condition (2.6a) guarantees that, assigning p to $(p, q) \in U \times M$ defines a map π: $E \to M$ that one uses to define (E, M, π) as a fiber space over M. (We leave the verification of details to the reader; of course, Steenrod [1] may be consulted.) One may think of the construction geometrically as "pasting together" pieces of $U \times M$ and $U \times M'$ to define E.

Finally, we come to the key notion of "*vector bundle*". Consider a fiber space (E, M, π). It is

---

[1]If $B \subset A \times A$ defines an equivalence relation on A, an *equivalence class* in A is a set of elements of A that are mutually equivalent, and that is contained in no larger such set. Then, each point of A belongs to precisely one such equivalence class. Define E, the *quotient* of A by the equivalence relation, as the collection of equivalence classes. There is then a "quotient map": $A \to E$, defined by assigning to each point of A the equivalence class to which it belongs.

called a *vector bundle* if:

a) For each $p \in M$, the fiber $\pi^{-1}(p)$ is a vector space (2.7)

b) The space M' used to define the local product structure is a vector space

c) The maps α_U: $E_U \to M'$ used to define the local product structure are *linear* transformations of the fibers of E onto the vector space M'.

Thus, conditions (2.8) imply that the transition maps $\{\beta_{U,U'}\}$ are maps of $U \cap U' \to$ (group of *linear* transformations on the vector space M'). Conversely, if this condition is satisfied, one readily verifies that Steenrod's "pasting together" construction, described above, leads to a vector bundle E. The key abstract idea is that the transition maps preserve the linear structure on the fiber M', hence this structure is "inherited" by the fibers of fiber spaces constructed by this "modelling" process.

<u>EXAMPLE</u>. Suppose that M is the unit sphere of R^n,

i.e. a point p of M is identified with an n-tuple $(x_1, \ldots, x_n)$ of real numbers such that:

$$x_1^2 + \ldots + x_n^2 = 1. \tag{2.8}$$

Now, intuitively, a vector $v = (v_1, \ldots, v_n) \in R^n$ is "tangent" to a point $p = (x_1, \ldots, x_n)$ of M if:

$$v_1 x_1 + \ldots + v_n x_n = 0 \tag{2.9}$$

The set of vector v satisfying (2.9) is, for fixed $(x_1, \ldots, x_n)$, an (n-1)-dimensional vector space, called the *tangent space to* M *at* p, which is denoted - in differential geometry - by M_p. The *tangent bundle* to M, denoted by T(M), is defined as follows:

> T(M) is the set of elements $(p, v) \in M \times R^n$ such that: $v \in M_p$, i.e. v is tangent to M at p. The projection π: $T(M) \to M$ is the map which assigns p to (p, v).

We must now exhibit the "local product" structure for this vector bundle. Now, M is a "differentiable manifold". In this case, it is given as "imbedded" as a "submanifold" of R^n. We can parameterize points of M by giving an open subset U of R^{n-1}, with points of U denoted by $t = (t_1, \dots, t_{n-1})$, together with a one-one map

$$t \to x(t),$$

such that

$$x(t_1)^2 + \dots + x(t_n)^2 = 1.$$

For example, such a parameterization can be obtained by singling out "x_1" (or any other coordinate axis of R^n), and defining

$$t \to (\sqrt{1-t_1^2 - \dots - t_{n-1}^2}, t_1, \dots, t_{n-1}) \tag{2.10}$$

For the parameterization (2.10), note that U consists of the points $t \in R^{n-1}$ such that $t_1^2 + \dots + t_{n-1}^2 < 1$. The points of M parameterized in this way, which can be identified with U as a subset of R^{n-1}, consists of the points $p = (x_1, \dots, x_n)$ with:

$x_1 > 0$.

We can then define α_U: $T(M)_U \to R^{n-1}$ as follows:

$$\alpha_U(p, v) = (v_2, \ldots, v_n).$$

Notice that, if $\alpha_U(p, v) = \alpha_U(p, v')$, then: $v_2' = v_2, \ldots, v_n'$. Using condition (2.9),

$$x_1(v_1 - v_1') = 0,$$

hence $v_1' = v_1$, since $x_1 > 0$, i.e. α_U is one-one on the fibers of $T(M)_U$. Since α_U is a linear mapping of finite dimensional vector spaces[1], it is onto, and defines an isomorphism of $T(M)_U$ with the product $U \times R^{n-1}$.

We can now let U' be the set of $p = (x_1, \ldots, x_{n-1}) \in M$ such that $x_1 < 0$, and apply a similiar construction. Of course, U and U' do

---

[1]If α: $V \to V'$ is a one-one, linear mapping of *finite dimensional* vector spaces, then a standard fact of linear algebra is that it is onto also, i.e. the inverse α^{-1} exists. Since α^{-1} is also linear, it is continuous (if V is a real or complex vector space) hence α is a homeomorphism.

not completely cover M; the points with $x_1 = 0$ escape. However, the construction can be iterated, with the other coordinate axes replacing the first; the resulting family of 2n open sets will cover M. This completes the demonstration that T(M) is a "vector bundle".

Of course, just because we were not successful *in this way* in exhibiting T(M) *globally* as a product does not settle the question whether it is possible some other way. In fact, this is a deep question in topology (See Atiyah [1] and Husemoller [1]); it is possibly *only* in the cases n = 1, 2, 4, 8.

Remarks. We have not been specific as to what sorts of vector spaces are allowed, i.e. as to what fields of scalars we want to allow. For our purposes, the real or complex numbers are enough. Accordingly, we speak of *real* or *complex* vector bundles according to whether the fibers of E are real or complex vector spaces, and the transition maps defined by the local products respect these structures.

3. TRANSFORMATION GROUPS, INTERTWINING MAPS, AND LINEAR GROUP ACTIONS ON VECTOR BUNDLES

We assume that the reader knows what groups are. They will be usually denoted by G, G',..., with individual elements denoted by g, g',... . The product law by which the "group" structure is defined is denoted by g_1g_2; the inverse by g^{-1}; and the identity by 1 or e, depending on the context.

As we mentioned in Section 2, a plentiful supply of groups is obtained from consideration of "transformation groups". Let M be a space. A *transformation* on M is a one-one, onto mapping $\alpha\colon M \to M$. The collection of these transformations is made into a group, as follows:

$\alpha_1\alpha_2$ is the transformation $p \to \alpha_1(\alpha_2(p))$ for $p \in M$.

α^{-1} is the inverse transformation to α.

<u>DEFINITION</u>. An "abstractly" given group G *acts as a transformation group on the space* M if there is given a map

$$h: G \to \text{(group of transformations on M)}$$

which is *also* a "homomorphism" in the sense of algebra, i.e.

$$h(g_1g_2) = h(g_1)h(g_2)$$

for $g_1, g_2 \in G$.

It is often convenient notationally to suppress explicit mention of h, and denote h(g) by g directly, so that, for $p \in M$, $g \in G$, gp denotes the transform of p by g (or h(g)).

As for any other homomorphism, the "kernel" of h, i.e. the set of $g \in G$ such that: $gp = p$ for all $p \in M$: forms an invariant subgroup G' of G. The quotient group G/G' (see below) then acts on M (since h "passes to the quotient" to define a one-one homomorphism on G/G'). G is said to *act trivially on* M if:

$$G = G'$$

G acts *effectively on* M if

$G' = (e)$, i.e. if no element of G

except the identity, e, acts as the identity map on M.

Associated with such a transformation group action on M are the notions of "orbits", "isotropy (or "little") subgroups", "coset spaces", etc. They have been discussed extensively in "Lie groups for physicists" [1], hence will only be briefly defined here.

> Given a point $p \in M$, the set of $g \in G$ which leave p fixed, i.e. such that $gp = p$, is a subgroup of G, denoted by G^p, called the *isotropy subgroup* of G at p. Given a point $p \in M$, the set of points of M of the form gp, for some $g \in G$, is called the *orbit* of G at p, denoted by Gp. The space whose "points" are the orbits of G on M is called the *orbit space* of G acting on M, denoted by $G\backslash M$. In other words, $G\backslash M$ is the quotient of M by the following equivalence relation: Points p', $p \in M$ are equivalent if there is a $g \in G$ such that $p' = gp$. Let G be a group, and let K be a subgroup. Translation by elements of K acting on G *on the right* defines an

action of K by transformations on G:

$$h(k)(g) = gk^{-1}$$

for $K \varepsilon K$, $g \varepsilon G$.

The orbits of K under this action are called the *right cosets* of K. The orbit space of G under this action is called the *coset space* or *homogeneous space* of G, with *isotropy subgroup* K, and is denoted by G/K. If G acts as a transformation group on a space M, and acts *transitively* on M, i.e. the orbit Gp of G at one point p of M fills up all of M, then the map $g \rightarrow gp$ of $G \rightarrow M$ passes to the quotient to define an identification of the coset space G/K with M, where K is the isotropy subgroup, G^p, of G at p.

With these concepts fixed, we pass to a new key idea:

DEFINITION. Let a group G act as transformation group on spaces M and M', and let ϕ be a map: $M \rightarrow M'$. ϕ is called an *intertwining map* for the action of G if:

$$\phi(gp) = g\phi(p)$$
for all $p \in M$, $g \in G$.

With these basic concepts in hand, turn to the case where π: $E \to M$ is a fiber space. A group G *acts on the fiber space* if G acts as a transformation group on E and M, with the map π intertwining the action.

Let us suppose that an action of G on the fiber space (E, M, π) is given. For $p \in M$, let E(p) denote the fiber $\pi^{-1}(p)$ over p. Then,

$$g(E(p)) = E(gp)$$
for $p \in M$, $g \in G$,

i.e. the action of G on E "permutes" the fibers. In particular, G^p, the isotropy subgroup of G at p, maps the fiber E(p) into itself.

In case E is a vector bundle, G is said to *act linearly* if:

g maps E(p) linearly onto E(gp),
for all $p \in M$, $g \in G$.

In particular, G^p acts linearly on the vector space

$E(p)$. This is called the *linear isotropy representation*.

If G acts on the fiber space (E, M, π), and acts transitively on M, it should be clear that the whole situation is determined up to isomorphism by the isotropy subgroup G^p at one point $p \in M$, and the action of G^p on $E(p)$. For, suppose that (E', M', π') is another such fiber space, with G acting on it, and with p' a point of M' such that: $G^p = G^{p'}$, $M = Gp$, $M' = Gp'$. As we have seen, M and M' can be considered as equal, since both may be identified with the coset space: $G/G^p = G/G^{p'}$. Then, an identification between $E(p)$ and $E(p')$ that intertwines the action of G^p can be extended to a map ϕ: $E \to E'$ that intertwines the action of G. To define this map, consider a point $x \in E$. Suppose that:

$$\pi(x) = gp.$$

Thus,

$$x' = g^{-1}x \in E(p).$$

Let:

$$\phi(x) = gx',$$

where we regard x' also as an element of E(p'), via the G-intertwining identification of E(p) with E(p'). One sees that $\phi(x)$ so defined is, in fact, independent of the g chosen to translate $\pi(x)$ back to p, hence defines a genuine map: $E \to E'$, which is the desired isomorphism between the fiber spaces.

We can apply these remarks, in particular, to the case where G^p acts as the identity on E(p). Then, E is isomorphic to the product $M \times E(p)$, with the action G defined as follows:

$$g(p, v) = gp, v) \qquad (3.1)$$

for $p \in M$, $v \in E(p)$, $g \in G$.

We can generalize (3.1) as follows. Suppose that E is isomorphic *as a fiber space* to the product $M \times E(p)$, while G acts on

Then,

$$g(p, v) = (gp, \alpha(p, g)(v)) \qquad (3.2)$$

where, for $p \in M$, $g \in G$, $\alpha(p, g)$ is a map: $E(p) \to E(p)$. Let us calculate the functional equation that α must satisfy.

$$(g_1 g_2)(p, v) = (g_1 g_2 p, \alpha(p, g_1 g_2)(v))$$
$$= g_1(g_2 p, \alpha(p, g_2)(v))$$
$$= (g_1 g_2 p, \alpha(g_2 p, g_1)\alpha(p, g_2)(v)).$$

Thus, we must have:

$$\alpha(p, g_1 g_2) = \alpha(g_2 p, g_1)\alpha(p, g_2) \tag{3.2}$$

for $p \in M$; $g_1, g_2 \in G$.

Conversely, these steps are reversible, and show that any map α: $M \times G \to$ (group of transformations on $E(p)$), satisfying the functional equation (3.2), defines an action of G on the product bundle $E = M \times E(p)$. Of course, if $E(p)$ is a vector space, and E is considered as a vector bundle, the condition that G act linearly is that $\alpha(p, g)$ act linearly on $E(p)$.

Notice from (3.2) that:

$$\alpha(p, g_1 g_2) = \alpha(p, g_1)\alpha(p, g_2)$$

for $g_1, g_2 \in G^p$.

Thus, $\alpha(p, G^p)$ is essentially the isotropy action of G^p on $E(p)$. It is again obvious from (3.2) that the isotropy action determines α uniquely, if G acts transitively on M.

4. DIFFERENTIABLE MANIFOLDS

Let us first review some general ideas concerning "coordinization" or "parameterization". Let M be a space of points. To "coordinatize" points of M means to assign a certain set of real numbers - say m of them - to each point of M, in such a way that to distinct points of M correspond distinct sets of numbers. Mathematically, this amounts to saying that a one-one map, denoted say by ϕ, is defined of $M \to R^m$, where R^m denotes the space of m-tuples of real numbers. (Thus, a point $x \in R^m$ may stand for an ordered m-tuple $(x_1,\dots,x_m)$ of real numbers $x_1,\dots, x_m$.)

For example, consider the possible initial states of motion of n particles moving in our

3-space. This may be considered abstractly, of course, since these words serve to construct a perfectly well-defined set. However, it is customary in physics to consider this set coordinatized in various ways. Let us consider the various physically interesting ways.

First, we may denote the initial position vectors of the particles by $\vec{r}_1, \ldots, \vec{r}_n$, where $\vec{r}_i$, $1 \leq i, j \leq n$, denote 3-vectors, i.e. points in R^3. Suppose that $(x_i, y_i, z_i) = \vec{r}_i$ are these Cartesian coordinates of these vectors.

Now, the future motion of the system of particles - when they are moving subject to Newton's laws of motion - are determined by their position vectors - described above - and their velocity vectors - denoted, say, by $\vec{v}_1, \ldots, \vec{v}_n$, with $\vec{v}_i = (u_i, v_i, w_i) \in R^3$. Thus, a possible "state" of the system is defined by a set

$$(x_1, y_1, z_1, u_1, v_1, w_1, x_2, y_2,$$
$$z_2, u_2, v_2, w_2, \ldots)$$

of 6n real numbers. This determines a coordinatization of the possible "states" of motion of the system.

Now, other "coordinatizations" are possible, and even important and useful for various physical purposes. For example, one might want to introduce momenta, "curvilinear coordinates", components with respect to moving coordinate systems, etc. The common abstract features to these constructions can be described as follows:

Let ϕ, ϕ' be maps: $M \to R^m$ which serve to coordinatize points of M. They are said to be *C^∞-equivalent* if there is a map α: $R^m \to R^m$ such that:

a) $\phi(p) = \alpha(\phi'(p))$
for all $p \in M$

b) α is a map that is C^∞ (read "infinitely differentiable")

c) The inverse map α^{-1}: $R^m \to R^m$ exists and is also C^∞.

Notice the role played by condition b) and c). They enable us to carry over the concepts of differential calculus to M, by referring via ϕ back to R^n. b) and c) guarantee that the concepts of "differentiability" a certain number of times, and such basic concepts as "differential equations", are really the same no matter which C^∞-equivalent coordinizations are chosen.

This can also be regarded group - theoretically. The set of α's satisfying a) and b) forms a group, the *group of diffeomorphisms* of R^n, denoted by: diff(R^n). Then, any concept that is "invariant" under the action of this group can be carried over to M in a way that does not depend on the choice of one particular coordinate system for M.

To develop the concepts of differential and integral calculus, on M in a way that is invariant under diff(R^n), it is been found that an algebraic terminology - based on the theory of "modules" - is most useful. Let F(M) denote the set of real - valued functions on M that are of differentiability class C^∞ when referred back via "coordinizations" to R^m, denoting a typical element of F(M) by $f: M \to R$. F(M) forms a (commutative, associative) algebra over the real numbers, since two such functions can be added, multiplied together, and multiplied by a scalar in an obvious way:

$$(f_1 + f_2)(p) = f_1(p) + f_2(p)$$

$$(f_1 f_2)(p) = f_1(p) f_2(p)$$

$$(cf_1)(p) = cf_1(p)$$

for f_1, $f_2 \in F(M)$, $p \in M$, $c \in R$.

DEFINITION. A linear map D: $F(M) \to F(M)$ is a *differential operator* if, when referred back to R^n via a coordinization, it is a differential operator in the usual sense. (Notice again that the property that the operator: $F(R^n) \to F(R^n)$ be a differential operator is invariant under diffeomorphisms of R^n.)

Let us make this more explicit. Suppose ϕ: $M \to R^m$ is coordinization of M. Thus, to each point $p \in M$ we associate m real numbers $(x_1, \ldots, x_m) = \phi(p)$, hence a function $f \in F(M)$ can be considered as a function $f'(x_1, \ldots, x_m)$ of m real variables. Then, we require that D have the following form:

$$D(f)(p) = \sum_{1 \leq i_1, \ldots, i_r \leq m} A_{i_1, \ldots, i_r}(x) \frac{\partial^r f'}{\partial x_{i_1}, \ldots, \partial x_{i_r}} + \ldots, \qquad (4.1)$$

where the A's are functions that are independent of f. The terms ... are terms of lower order in the partial derivatives.

Now, if we change the coordinization by $\alpha \in \mathrm{diff}(R^m)$, of the form:

$$\alpha(x) = x' = (x_i'(x)),$$

the specific form of the right hand side will change, in accordance with the chain-rule for partial derivatives:

$$\frac{\partial}{\partial x_i} = \left(\frac{\partial x_j'}{\partial x_i}\right)\frac{\partial}{\partial x_j'} \tag{4.2}$$

(From now on, we adopt the following range of indices, and the summation convention:

$$1 \leq i,\ j,\ i_1,\ i_2, \ldots, \leq m).$$

However, notice that the fact that D is a differential operator, i.e. that it can be written in the *form* (4.1) for *some* choice of the A's, is independent of the choice of coordinization. Notice that the integer r appearing in (4.1) is also independent; let us call it the *order* of the differential operator. The set of differential operators whose order is at most r is denoted by:

$$\underset{\sim}{D}^r(M)$$

The space of differential operators of any order is denoted by:

$$\underset{\sim}{D}(M).$$

We can now describe various algebraic operations on differential operators:

a) *Addition:*

For $D_1, D_2 \in \underset{\sim}{D}(M)$, $f \in F(M)$,

$$(D_1 + D_2)(f) = D_1(f) + D_2(f)$$

b) *Product:*

$$(D_1D_2)(f) = D_1(D_2(f))$$

c) *F(M)-multiplication:*

For $f_1 \in F(M)$, $D \in \underset{\sim}{D}(M)$, $f \in F(M)$,

$$(f_1D)(f) = f_1D(f)$$

d) *Anticommutator or Jordan bracket*

$$[D_1, D_2]_+ = D_1D_2 + D_2D_1$$

These algebraic operations can be classified under the jargon of the theory of "algebras". We will review this jargon in the next section.

There is another way of making explicit what is meant by a differential operator that we will briefly describe, using the idea of the theory of "moving frames". Suppose ϕ: $M \to R^m$ is a coordinization of M. As above, to each $f \in F(M)$ we can assign a function $f'(x)$ of an m-vector $x = (x_1, \ldots, x_m)$, i.e.

$$f(p) = f'(\phi(p)) \quad \text{for all} \quad p \in M.$$

This correspondence $F(M) \to F(R^m)$ defines an isomorphism between the two algebras. Let $\frac{\partial}{\partial x_i} \in D^1(M)$ be the first-order differential operators defined by differentiation along the coordinate axes of R^m. Explicitly,

$$\frac{\partial}{\partial x_1}(f)(p) = \lim_{\varepsilon \to 0} \frac{1}{\varepsilon}[f(\phi(p)+(\varepsilon, 0,\ldots,0))-f(\phi(p))]$$

$$\frac{\partial}{\partial x_2}(f)(p) = \lim_{\varepsilon \to o} \frac{1}{\varepsilon} [f(\phi(p)$$

$$+ (0, \varepsilon, 0, \ldots, 0)) - f(\phi(p))$$

and so forth.

For $f_1 \varepsilon F(M)$, let D_{f_1} be the operation $f \to f_1 f$ of multiplicatively f_1. Then a *differential operator on* M is a linear mapping: $F(M) \to F(M)$ which can be written in terms of a finite number of iterations of the operators $\frac{\partial}{\partial x_i}$, D_{f_1}, via the operations a) - c) described above.

So far, for the sake of simplicity we have been dealing with spaces M which can be coordinitized by one-one maps between M and open subsets of R^m. This is not sufficient to cover all cases of interest; for example, it forces M to be non-compact, as a topological space (if one requires ϕ to be a homeomorphism with respect to a topology given in advance for M, which is a reasonable condition, of course). For example, the case treated above, where M is the unit sphere in R^{m+1}, would not be included since it cannot be covered by a single coordinization.

Luckily, this problem can be surmounted by the remark that all of the concepts we will need are "local" in nature, hence apply to a general sort of "differentiable manifold", described as follows:

<u>DEFINITION</u>. Let M be a topological space. M is a *differentiable manifold* if there are a countable sequence M_1, M_2,... of open subsets of M such that:

a) Each open subset of M_α, $\alpha = 1, 2, 3, \ldots$, carries a homeomorphism ϕ_α between M_α and an open subset of R^m, enabling one to coordinatize points of M_α by real numbers, as described above, and to define the algebra $F(M_\alpha)$ of C^∞, real-valued functions on M_α.

b) Each point of M is contained in at least one of the sets M_1, M_2,... (possibly many, of course), i.e. $\{M_\alpha\}$ forms an "open covering" of M.

c) if M_α and M_β intersect, then the restriction of the functions in $F(M_\alpha)$ and $F(M_\beta)$ to $M_\alpha \cap M_\beta$ agree, i.e. if a real-valued function f: $M_\alpha \to R$ is C^∞ when referred back to R^m via ϕ_α, its restriction to $M_\alpha \cap M_\beta$ is C^∞ when referred back to R^m via ϕ_β, for all α, β.

Condition c) may be reformulated by saying that the notion of something being C^∞ is independent of the coordinization chosen. We can also reformulate c) in terms of mappings. ${\phi_\alpha}^{-1}$ maps $\phi_\alpha(M_\alpha)$ isomorphically onto M_α, hence $\phi_\beta{\phi_\alpha}^{-1}$ maps $\phi_\alpha(M_\alpha \cap M_\beta) \subset R^m$ in a one-one way onto the open subset $\phi_\beta(M_\alpha \cap M_\beta)$ of R^m. Condition c) is equivalent to requiring that this "transition mapping" be - for all α, β of course - an infinitely differentiable mapping as this term is applied to mappings between open subsets of Euclidean space.

Now, we can free ourselves from explicit mention of these coordinization maps ϕ_α. Let us say that a real-valued function f: $M \to R$ is C^∞ if its restriction to each M_α belongs to $F(M_\alpha)$, as defined above. Denote the algebra of such f's by F(M).

Consider a map ϕ: $M \to M'$ between two such manifolds. For $f' \in M'$, let $\phi^*(f')$ denote the following real-valued function on M:

$$\phi^*(f')(p) = f'(\phi(p)).$$

This map ϕ^* is called the "pull-back" or "dual" mapping to ϕ. We say that the map ϕ is of *differentiability class* C^∞ if:

$$\phi^*(f') \in F(M)$$

for all $f' \in F(M')$.

In this work, we will always deal with C^∞ *maps, unless specifically mentioned otherwise.*

Thus, ϕ^* is an algebra homeomorphism: $F(M') \to F(M)$. (See next section for the algebraic terminology used here.) ϕ is called a diffeomorphism between M and M' if the inverse map ϕ^{-1}: $M' \to M$ exists, and is C^∞. (Exercise: Prove that a map ϕ: $R^m \to R^{m'}$ is C^∞ in this sense if and only if it is C^∞ in the usual sense, i.e. when expressed in terms of real-variables of R^m and $R^{m'}$).

Suppose now that $\{M_\alpha\}$, $\{M_\alpha'\}$ are two open covers of M that lead to manifold structures for M. We say they lead to *equivalent manifold structures* for M if the identity map: $M \to M$ is a diffeomorphism, as defined in the last paragraph. (It should be pointed out that there is a reason for this caution: It has been discovered by J. Milnor - and this is one of the great triumphs of algebraic topology in the last fifteen years - that there may be two inequivalent manifold structures for the same topological space, e.g. the seven-dimensional sphere in Euclidean space.)

5. ALGEBRAS AND MODULES

Let $\underset{\sim}{A}$ be a vector space, say over the real numbers as "ground field". Denote typical elements of $\underset{\sim}{A}$ by $A_1, A_2, \ldots$. An *algebraic structure* for $\underset{\sim}{A}$ is defined by a bilinear map: $\underset{\sim}{A} \times \underset{\sim}{A} \to \underset{\sim}{A}$. For the moment, denote the image of (A_1, A_2) under this map by: A_1A_2. Then, the "bilinearity" just means the "distributive law" between addition and multiplication:

$$A_1(A_2 + A_3) = A_1A_2 + A_1A_3$$

$$(A_1 + A_2)A_3 = A_1A_3 + A_2A_3.$$

<u>DEFINITION</u>. $\underset{\sim}{A}$ is an *associative algebra* if

$$A_1(A_2A_3) = (A_1A_2)A_3$$

for $A_1, A_2, A_3 \varepsilon \underset{\sim}{A}$.

The typical example of an associative algebra is the space of linear operators on a vector space, with the "product" A_1A_2 just the composition of operators.

DEFINITION. $\underset{\sim}{A}$ is a *Lie algebra* if:

a) $A_1(A_2A_3) = (A_1A_2)A_3 + A_2(A_1A_3)$. (5.1)

b) $A_1A_2 = - A_2A_1$

for $A_1, A_2, A_3 \varepsilon \underset{\sim}{A}$.

Condition (5.1a) is called the *Jacobi identity*. It is customary to denote the product A_1A_2 in this Lie case by: $[A_1, A_2]$. A class of examples of Lie algebras may be constructed by starting off with an associative algebra $\underset{\sim}{A}$, with its "associative" product denoted by A_1A_2, and *define* its Lie product as the commutator:

$$[A_1, A_2] = A_1A_2 - A_2A_1.$$

We now turn to the concepts associated with mappings. A mapping ϕ: $\underset{\sim}{A} \rightarrow \underset{\sim}{A}'$ between algebras is called a *homomorphism* if:

a) ϕ is a linear map

b) $\phi(A_1A_2) = \phi(A_1)\phi(A_2)$

for $A_1, A_2 \in \underset{\sim}{A}$.

Consider such a map. Let:

$$\underset{\sim}{B} = \phi^{-1}(0) = \text{kernel of } \phi.$$

Then,

$$\phi(AB) = \phi(A)\phi(B) = 0$$

$$\phi(BA) = \phi(B)\phi(A) = 0$$

$$\text{for} \quad A \in \underset{\sim}{A},\ B \in \underset{\sim}{B}.$$

Thus,

$$\underset{\sim}{A}\underset{\sim}{B} \subset \underset{\sim}{B};\ \underset{\sim}{B}\underset{\sim}{A} \subset \underset{\sim}{A} \tag{5.2}$$

A subspace $\underset{\sim}{B}$ satisfying (5.2) is called an *ideal* of $\underset{\sim}{A}$. Conversely, if $\underset{\sim}{B}$ is such an ideal, and if $\underset{\sim}{A}'$ is *defined* as the vector space quotient $\underset{\sim}{A}/\underset{\sim}{B}$, as the quotient map, then the algebra structure on $\underset{\sim}{A}$ "passes to the quotient" to define an algebra structure on $\underset{\sim}{A}'$ such that ϕ is a homomorphism.

A more inclusive notion than ideal is that of "subalgebra". A linear subspace $\underset{\sim}{B} \subset \underset{\sim}{A}$ is a *subalgebra* if:

$$\underset{\sim}{B}\underset{\sim}{B} \subset \underset{\sim}{B}.$$

Subalgebras arise as image sets of homomorphism: If ϕ: $\underset{\sim}{A} \to \underset{\sim}{A}'$ is an algebra homomorphism, and if $\underset{\sim}{B} = \phi(A)$, then $\underset{\sim}{B}$ is a subalgebra.

Now, turn to the consideration of "modules". Let F be a commutative, associative algebra. A vector space $\underset{\sim}{A}$ is a *module* over F if there is a bilinear map: $F \times \underset{\sim}{A} \to \underset{\sim}{A}$, denoted by $(f, A) \to fA$, such that:

$$f_1(f_2A) = (f_1f_2)A$$

$$\text{for} \quad f_1, f_2 \in F, A \in \underset{\sim}{A}.$$

A linear map ϕ: $\underset{\sim}{A} \to \underset{\sim}{A}'$ between F-modules is said to be *F-linear* if:

$$\phi(fA) = f\phi(A)$$

for $f \in F$, $A \in \underset{\sim}{A}$.

A set of elements $A_1, \ldots, A_n$ of an F-module $\underset{\sim}{A}$ is said to form a *module basis* for $\underset{\sim}{A}$ if each element $A \in \underset{\sim}{A}$ can be written *in a unique manner* in the form:

$$A = f_1 A_1 + \ldots + f_n A_n$$

with a set $(f_1, \ldots, f_n)$ of elements of F.

A module that admits such a basis is called a *free module*. (Not all modules are free-although for most practical purposes in differential geometry the study of free modules is sufficient.)

Modules may be constructed algebraically using the tensor product. Start off with a vector space V. Set:

$$\underset{\sim}{A} = F \otimes V.$$

i.e. $\underset{\sim}{A}$ is the tensor product of *the real vector spaces* F and V. (We assume, of course, that the reader knows the tensor-product construction.)

Define $\underset{\sim}{A}$ as an F-module as follows:

$$f(f_1 \otimes v) = (ff_1) \otimes v$$

for $f, f_1 \in F$, $v \in V$.

6. DIFFERENTIAL OPERATORS ON VECTOR BUNDLES

From now on, all "spaces" will be differentiable manifolds, and all maps will be differentiable manifolds, unless mentioned otherwise.

Let (E, M, π) be a vector bundle. A *cross-section*, typically denoted by Ψ, is a map: $M \to E$ such that:

$$\Psi(p) \in E(p) \quad \text{for all} \quad p \in M. \tag{6.1}$$

($E(p)$ denotes the fiber $\pi^{-1}(p)$ of the fiber space). Since by the definition of vector bundle, $E(p)$ is a vector space (say, over the real or complex numbers), two cross-sections Ψ_1, Ψ_2 may be added point-wise:

$$(\Psi_1 + \Psi_2)(p) = \Psi_1(p) + \Psi_2(p)$$

for $p \in M$.

Thus, the space of cross-sections, denoted by $\Gamma(E)$, forms a vector space. In "Lie groups for physicists" [1] we have extensively discussed the role this vector space plays in the representation theory of Lie groups. One of the principal aims in this book is to describe the role it plays in quantum field theory.

In addition to its vector space structure, $\Gamma(E)$ has a "module" structure that plays a key role in differential geometry. Let $F(M)$ denote the algebra (over the real numbers) of real-valued, C^∞ functions on M. $\Psi \in \Gamma(E)$ can be multiplied by a $f \in F(M)$:

$$(f\Psi)(p) = f(p)\Psi(p) \quad \text{for} \quad p \in M.$$

The resulting product: $F(M) \times \Gamma(E) \to \Gamma(E)$ defines $\Gamma(E)$ as an $F(M)$-module. (See Section 5 for the algebraic terminology to be used here.) We will use this module structure in a key manner to define the concept of "differential operator".

One approach to the definition of a differential operator proceeds via the local product structure assumed for the bundles. Suppose that

(E, M, π) and (E', M, π') are two vector bundles over the same base space M. Let p_0 be a fixed point of M, and suppose that E and E' are isomorphic to the products $M \times E(p_0)$ and $M \times E'(p_0)$. Thus, a point of E can be identified with a pair (p, v), with $p \in M$, $v \in E(p)$, and

$$\pi(p, v) = p.$$

If $\Psi \in \Gamma(E)$, then

$$p = \pi\Psi(p) \quad \text{for} \quad p \in M,$$

by the definition (6.1) of cross-section, hence: $\Psi(p)$ is of the form:

$$\psi(p) = (p, \phi(p)),$$

with $\phi(p) \in E(p_0)$. Thus, $p \to \phi(p)$ defines a map: $M \to E(p_0)$. Conversely, any such map arises in this way from a cross-section. Thus, we see that:

> There is a one-one correspondence between cross-sections and maps of M into the fiber.

Now, chosing bases for the vector spaces $E(p_o)$ and $E'(p_o)$, we see that cross-section of E and E' may be identified with sets of real-valued functions defined on M, called the *components* of the cross-sections.

DEFINITION. If E and E' are product vector bundles, then an R-linear map D: $\Gamma(E) \to \Gamma(E')$ is a *differential operator* if it acts on the components of cross-sections by differentiating them.

One sees readily that this property is independent of the chosen product structure, hence is an "intrinsic" property of the vector bundles. We shall describe more explicitly below how the differential operators acting on the components may be described.

Our next task is to show how a differential operator may be defined for bundles that are not products.

DEFINITION. A linear map D: $\Gamma(E) \to \Gamma(E')$ is said to be *local* if the following condition is satisfied:

For every open subset M_o of M, and every

$\Psi \in \Gamma(E)$ that vanishes at each point of M_o, $D(\Psi)$ also vanishes at each point of M_o.

As its name indicates, this property enables us to "localize" D, in the following way. Let M_o be an open subset of M. Let $E_o = \pi^{-1}(M_o)$, $E_o' = \pi^{-1}(M_o)$ be the vector bundles obtained by restricting to M_o. If D is local operator, one can then define

$$D_o: \ \Gamma(E_o) \to \Gamma_o(E_o')$$

in the following way:

Given $\Psi_o \in \Gamma(E)$, extend[1] Ψ_o to a cross-section Ψ, and then define:

$$D_o(\Psi) = D(\Psi) \text{ restricted to } M_o. \qquad (6.2)$$

The "locality" property of D then guarantees that D_o is independent of the extension

[1]For example, using a partition of unity. See Loomis and Sternberg [1] or Spivak [1].

DEFINITION. A linear operator D: $\Gamma(E) \to \Gamma(E')$ is a *differential operator from* E *to* E' if:

a) D is local

b) For every open subset M_o of M over which the bundles E, E' are products, the "localization" operator D_o: $\Gamma(E_o) \to \Gamma(E_o')$ defined by (6.2) is a differential operator, as defined previously, i.e. acts by differentiation on the components.

If the fibers of E and E' are finite dimensional vector spaces, one can in fact, prove that condition b) is superfluous, i.e. it follows from a). However, this is technically difficult to prove (see Peetre [1]), and will not be used to our work.

Let us now reformulate things in terms of the F(M)-module structure of $\Gamma(E)$. In Section 5, we have defined the concept of "basis" of a module. Let (Ψ_a); $1 \le a, b, \dots, \le n$; $(\Psi'_{a'})$; $1 \le a', b', \dots, \le n'$ (summation convention in force) be for the F(M)-modules $\Gamma(E)$ and $\Gamma(E')$. Let D: $\Gamma(E) \to \Gamma(E')$ be an R-linear operator. Thus, we have:

$$D(f\Psi_a) = \Delta_{aa'}(f)\Psi'_{a'} , \qquad (6.3)$$

for all $f \varepsilon F(M)$, where $(\Delta_{aa'})$ are R-linear operators:

$$F(M) \to F(M).$$

Then, one sees readily that D is a differential operator in the sense defined above if and only if the $\Delta_{aa'}$ are differential operators on M. These operators are called the *components* of D with respect to the "moving frames" (Ψ_a), $(\Psi'_{a'})$. The maximal order of the differential operators $\Delta_{aa'}$ is called the *order* of D.

$\underset{\sim}{D}(E, E')$ will denote the space of differential operators from E to E'. $\underset{\sim}{D}^r(E, E')$ will denote the space of those operators that are of order no greater than r. If $E = E'$, this will be abbreviated to:

$$\underset{\sim}{D}(E), \underset{\sim}{D}^r(E').$$

$\underset{\sim}{D}(E, E')$ and $\underset{\sim}{D}^r(E, E')$ are themselves F(M)-modules, since differential operators can be multiplied by functions:

$$(fD)(\Psi) = fD(\Psi)$$

for $f \in F(M)$, $D \in \underset{\sim}{D}(E, E')$.

In differential geometry, the "homogeneous, first order, differential operators" play a special role (because they serve as infinitesimal generators of one parameter transformation groups), hence will be assigned special notation.

Let V(M) denote the space of differential operators $X\colon\ F(M) \to F(M)$ such that:

$$X(f_1 f_2) = X(f_1)f_2 + f_1 X(f_2) \qquad (6.4)$$

for $f_1, f_2 \in F(M)$.

Condition (6.4) means that X is a *derivation* of the algebra F(M). They form a Lie algebra under commutor:

$$[X_1, X_2] = X_1 X_2 - X_2 X_1 \qquad (6.5)$$

In differential geometry, the elements of V(M) are called *vector fields* on M.

Let (E, M, π) be a vector bundle over M. Denote by V(E, M) the operators $D \in \underset{\sim}{D}^1(E)$ with the following property:

There is an $X \in V(M)$ such that

$$D(f\Psi) = fD(\Psi) + X(f)\Psi \qquad (6.6)$$

for $f \in F(M)$, $\Psi \in \Gamma(E)$

Again, one sees that $V(E, M)$ forms a Lie algebra under commutator. The assignment $D \to X$ defines a Lie algebra homeomorphism of $V(E, M) \to V(M)$.

We can briefly indicate the geometric genesis of these operators. Suppose that G is a group, that acts as a transformation group on E and M, such that:

a) π intertwines the action of G on E and M

b) For $g \in G$, $p \in M$, g maps $E(p)$ *linearly* onto $E(gp)$.

(Recall that these two conditions together say that G *acts linearly on the vector bundle.*) Suppose that $t \to g(t)$ is a one-parameter subgroup of G, i.e.

$$g(t_1 + t_2) = g(t_1)g(t_2)$$

for $t_1, t_2 \in R$.

For $\Psi \in \Gamma(E)$, $g \in G$, define

$$g(\Psi) \in \Gamma(E) \text{ so that:}$$
$$g(\Psi)(p) = g(\Psi(g^{-1}p)).$$

Then, G so defined acts linearly on $\Gamma(E)$, i.e. defines a representation of G by linear operators on $\Gamma(E)$. (In "Lie groups for physicists", it is shown how this ties in with the theory of "induced representations".) Then, define:

$$D(\Psi) = \frac{\partial}{\partial t} g(t)(\Psi)\Big/_{t=0} \tag{6.7}$$

$$X(f) = \frac{\partial}{\partial t} g(t)(f)\Big/_{t=0} \tag{6.8}$$

for $\Psi \in \Gamma(E)$, $f \in F(M)$.

It is readily verified that (6.6) is satisfied. If G is a Lie group, the set of one-parameter subgroups may be identified with its Lie algebra $\underset{\sim}{G}$. Then, one can show that (6.7) and (6.8) define homeomorphisms of $\underset{\sim}{G} \to V(E, M)$ and $\underset{\sim}{G} \to V(M)$.

7. JET-BUNDLES ASSOCIATED WITH MODULES

Let M be a manifold, with F(M) its algebra of

C^∞, real-valued functions. Consider an F(M)-module, Γ. (For the moment, we will work abstractly - the relevance to Γ's that arise from cross-sections of vector bundles will be explained further on.)

Let p be a point of M. Denote by $F^p(M)$ subspace of $f \in F(M)$ such that:

$$f(p) = 0.$$

Denote by Γ^p the subspace of Γ consisting of the linear combinations of elements of $F^p(M)$; i.e. the space of elements $\Psi \in \Gamma$ that can be written in the form:

$$\Psi = f_1\Psi_1 + \ldots + f_s\Psi_s, \tag{7.1}$$

with $f_1(p) = 0 = \ldots = f_s(p)$, $\Psi_1, \ldots, \Psi_s \in \Gamma$. The Ψ of form (7.1) may be thought of as those that *vanish to the first order at* p. Define:

$$E(p) = \Gamma/\Gamma^p.$$

Then, E(p) is a vector space over the real numbers. Define E as the set of ordered pairs (p, v), with $v \in E(p)$. Define a map π: E $\to$ M as follows:

$$\pi(p, v) = p.$$

Then, the triple (E, M, π) has obviously some of the properties of a vector bundle. Indeed, if we start off with Γ equal to the space of cross-sections of a vector bundle (E', M, π'), we can show that (E, M, π) is isomorphic as a vector bundle to (E', M', π'), with the isomorphism ϕ: $E \to E'$ defined as follows:

$$\phi(p, v) = \Psi(p),$$

where Ψ is any element of $\Gamma(E')$ which goes over, under the quotient map:

$$\Gamma(E) \to \Gamma(E)/\Gamma^p(E) = E(p)$$

into v.

However, reformulating things in this way suggests a natural generalization: Divide out Γ by those elements that vanish to the "first", "second", "..." ... order at p. Let us make this more precise.

Having defined Γ^p - the objects vanishing "to the zeroth order" at p - let us define $\Gamma_1{}^p$ -

those vanishing to the "first order" at p, as follows:

> $\Gamma_1{}^p$ is the space of linear combinations of elements of $F^p(M)\Gamma^p$
>
> $\Gamma_2{}^p$ is the space of linear combinations of elements of $F^p(M)\Gamma_1{}^p$.

Continue by induction: $\Gamma_r{}^p$ is the space of linear combinations of $F^p(M)\Gamma_{r-1}{}^p$.

Now set:

$$J^r(E)(p) = \Gamma/\Gamma_{r-1}{}^p \tag{7.2}$$

$J^r(E)$ = space of ordered pairs (p, v), with $p \in M$, $v \in J^r(E)(p)$.

π^r: $J^r(E) \to M$, is defined as $\pi^r(p, v) = p$.

Thus, again $(J^r(E), M, \pi^r)$ has the algebraic properties of a "vector bundle", i.e. π^r is a map of $J^r(E)$ onto M, whose fibers are vector spaces. One can prove (see Palais [1, 2]) that, if $\Gamma = \Gamma(E)$, where E is a genuine vector bundle, that $J^r(E)$ also satisfies the additional topological condition needed to define a vector bundle. This suggests that we adapt our terminology - in speaking of $J^r(E)$ - to the case where it is a vector bundle.

$J^r(E)$ is called the bundle of *r-th order jets* associated with the module Γ, or the bundle E, defined by: $E(p) = \Gamma/\Gamma^o$.

These "jet-bundles" are vital for the study of differential operators. Let us start off with the study of "homogeneous, first order operators". Suppose D is a linear map: $\Gamma \to \Gamma$, and X is a linear map: $F(M) \to F(M)$ of the type studied in Section 6, i.e.

$$D(f\Psi) = X(f)\Psi + fD(\Psi)$$
$$\text{for} \quad \Psi \in \Gamma,\ f \in F(M). \tag{7.3}$$
$$X(f_1 f_2) = X(f_1)f_2 + f_1 X(f_2)$$
$$\text{for} \quad f_1,\ f_2 \in F(M).$$

Consider $\Psi \in \Gamma$, $f_1, \ldots, f_r \in \Gamma^p$. Then,

$$f_1 \ldots f_r \Psi \in \Gamma_{r-1}{}^p.$$

Then,

$$D(f_1 \ldots f_r) = X(f_1)f_2 \ldots X_r \Psi$$
$$+ f_1X(f_2)f_3 \ldots X_r \Psi$$
$$+ \ldots + f_1 \ldots f_rD(\Psi),$$

which belongs to $\Gamma_{r-2}{}^p$. Then, we have:

$$D(\Gamma_{r-1}{}^p) \subset \Gamma_{r-2}{}^p. \qquad (7.4)$$

In particular, for each p, r, D passes to the quotient to define a linear map

$$\sigma_r(D)(p): J^r(E)(p) \to J^{r-1}(E)(p). \qquad (7.5)$$

As p varies, $\sigma_r(D)$ defines a map: $J^r(E) \to J^{r-1}(E)$ that is linear on the fibers of the vector bundles.

Now, suppose that D is of the form D_1D_2, where D_1, D_2 are operators: $\Gamma \to \Gamma$ of form (7.3). Then, iterating (7.4), we see that:

$$D(\Gamma_{r-1}{}^p) \subset \Gamma_{r-3}{}^p,$$

hence D passes to the quotient to define a map:

$$\sigma_r(E)(p): \quad J^r(E)(p) \to J^{r-2}(E)(p),$$

and, as p varies, a linear bundle map: $J^r(E) \to J^{r-2}(E)$. Now, the key property enabling us to define this map $\sigma_r(D)$ is that D is a second order differential operator. Let us then generalize as follows:

Suppose E, E' are vector bundles over M, and D: $\Gamma(E) \to \Gamma(E')$ is a differential operator of order s, i.e. $D \in \underset{\sim}{D}^s(E, E')$. Then, for $p \in M$,

$$D(\Gamma_{r-1}{}^p(E)) \subset \Gamma_{r-1-s}{}^p(E),$$

hence D passes to the quotient to define a linear map $\sigma_r(D)$: $J^r(E)(p) \to J^{r-s}(E')(p)$. As p varies, $\sigma_r(D)$ defines a linear bundle map:

$$\sigma_r(D): \quad J^r(E) \to J^{r-s}(E').$$

We can make this more explicit using a module basis for $\Gamma(E)$ and $\Gamma(E')$, and a coordinate system for the manifold M. Suppose that (Ψ_a), $(\Psi'_{a'})$; $1 \leq a, b, \ldots, \leq n$; $1 \leq a', b', \ldots, \leq n'$; are F(M)-

module bases, and M is coordinatized by identifying a point p with an m-vector (x_i), $1 \leq i; j, \ldots, \leq m$, of real numbers. Suppose that:

$$D(f\Psi_a) = \Delta_{aa'}(f)\Psi_{a'},$$

for $f \in F(M)$,

where:

$$\Delta_{aa'} = A_{aa'}^{i_1 \ldots i_s} \partial_{i_1} \ldots \partial_{i_s} + \ldots \qquad (7.6)$$

with: $\partial_i = \dfrac{\partial}{\partial x_i}$. (The terms ... are of lower order.)

Given $f \in F(M)$, we can consider its Taylor expansion about the point p:

$$f = f(p) + f_i(p)(x_i - x_i(p))$$
$$+ f_{i_1 i_2}(p)(x_i - x_j(p))(x_j - x_j(p)) + \ldots .$$

Suppose, for simplicity, that

$$x_i(p) = 0.$$

Then, a basis for $\Gamma(E)/\Gamma_{r-1}{}^{p}(E)$ consists of the elements:

$$\Psi_a, x_i\Psi_a, \ldots, x_{i_1} \ldots x_{i_{r-1}}\Psi_a.$$

Similarly, a basis for $\Gamma(E')/\Gamma_{r-1-s}{}^{p}(E)$ consists of $\Psi'_{a'}, x_{i_1}\Psi'_{a'}, \ldots, x_{i_1} \ldots x_{i_{r-1-s}}\Psi'_{a'}$. Thus, using (7.6), we can compute the matrix of $\sigma_r(D(p))$ with respect to these bases, in terms of the coefficients A occurring in (7.6). The formulas are complicated, and there is no point in writing them out in general here. We shall only do two special cases:

a) $s = 1, r = 1$:

$$\Delta_{aa'} = A_{aa'}{}^{i}\partial_i + A_{aa'}$$

$$D(\Psi_a) = A_{aa'}(p)\Psi_{a'}$$

$$D(x_i\Psi_a) = A_{aa'}{}^{i}(p)\Psi_{a'} + \ldots$$

(The terms ... fall out when the quotient by $\Gamma^p(E)$ is taken)

b) $s = 2, r = 1$

$$\Delta_{aa'} = A_{aa'}{}^{ij}\,\Psi_{a'} + A_{aa'}{}^{i}\partial_i + A_{aa'}$$

$$D(\Psi_a) = A_{aa'}(p)\Psi_{a'}$$

$$D(x_i\Psi_a) = \Delta_{aa'}(x_i)\Psi_{a'} = A_{aa'}{}^{i}(p)\Psi_a + \ldots$$

$$D(x_i x_j \Psi_a) = \Delta_{aa'}(x_i x_j)\Psi_{a'}$$

$$= (A_{aa'}{}^{ij}(p) + \ldots\ )$$

Let us return to the general theory. If $r' < r$, then $\Gamma_{r'-1}{}^{p}(E) \supset \Gamma_{r-1}{}^{p}(E)$, since if a $\Psi \in \Gamma(E)$ vanishes to order $(r-1)$, then it also vanishes to order $(r'-1)$. Thus, there is a quotient linear mapping:

$$J^r(E)(p) \to J^{r'}(E)(p) \tag{7.7}$$

As p varies, this induces a linear bundle map: $J^r(E) \to J^{r'}(E)$. If $D \in \underset{\sim}{D}^s(E, E')$, then obviously we have a "commutative diagram":

$$\begin{array}{lccc} \sigma_r(D): & J^r(E) & \to & J^{r-s}(E') \\ & \downarrow & & \downarrow \\ \sigma_{r'}(D): & J^{r'}(E) & \to & J^{r'-s}(E'). \end{array}$$

(By a "commutative diagram" (of maps), we mean that composing the maps in any order leads to the same result.)

In particular, let us apply this to the case: $r' = r-1$. Set:

$J_h{}^r(E)$ = kernel of the map defined in (7.7) of

$J^r(E)(p) \to J^{r-1}(E)(p)$.

As the notation indicates, the elements of $J_h{}^r(E)$ may be thought of as the "r-th order homogeneous jets". Then, if $D \in \underset{\sim}{D}^s(E, E')$, $\sigma_r(D)$ maps $J_h{}^r(E)$ linearly into $J_h{}^{r-s}(E')$. For $r = s$, this linear map is called the *symbol* of the differential operator D. One sees that the symbol is determined by the terms $A_{aa'}^{i_1,\ldots,i_s}$ of order s in the expression (7.6) for D in terms of "moving frames".

Finally, we might remark that the theory of "jets" was created by C. Ehresmann[1] in the 1950's

---

[1]When names are presented in this way, without references, it is expected that the reader will consult Mathematical Reviews.

to provide a general formalism for differential geometric problems. Essentially, it enables one to handle partial derivatives in a very natural geometric way - although it is initially a confusing formalism to learn. It has wide ramifications in topology and partial differential equations; for example, consult the work of R. Thom, D. C. Spencer, A. Sternberg, V. Guillemin, H. Goldschmidt, and M. Kuranishi.

We have now completed our exposition of the essential elements of the theory of vector bundles, differential operators, and jet-bundles. Of course, so far we have only been concerned with formalism and notations. However, perhaps unfortunately, such details are critical in the "modern" theory of differential operators. To find some connection with physics, we now turn to quantum field theory and the calculus of variations.

CHAPTER II

JET BUNDLES AND THE CLASSICAL CALCULUS OF VARIATIONS

The aim of this chapter is to show how "fiber bundle" language provides a convenient way to describe some of the features of the calculus of variations formalism associated with "classical" fields. Many of these features are also important in quantum field theory.

1. JET BUNDLES ASSOCIATED WITH FIBER SPACES

In Chapter 1, we have described a fairly complete formalism for the description of "jets" of cross-sections of vector bundles, and described its relation to the theory of linear differential operators. The theory we are about to describe amounts to an attempt to generalize some of the ideas to describe a theory of "non-linear differential operators". However, since the theory of these objects is in a much less complete stage, we will not attempt to be very general, but will remain fairly close to the machinery needed for the applications to the calculus of variations.

Suppose that E and M are differentiable manifolds, and that π: $E \to M$ is a projective mapping defining a "local product fiber space", as described in Chapter 1. $\Gamma(E)$ denotes its space of sections, i.e. the space of maps Ψ: $M \to E$ such that:

$$\Psi(p) \in E(p) \quad \text{for all} \quad p \in M$$

(Recall that $E(p) = \pi^{-1}(p)$ denotes the fiber above the point p.)

Now, E is not necessarily a vector bundle, so $\Gamma(E)$ does not have a natural vector space structure which we can use to define the "jets". Instead, we will proceed directly to define an equivalence relation, whose quotient will define "jets".

Given Ψ, $\Psi' \in \Gamma(E)$, and a point $p \in M$, what is wanted is some way of saying that "Ψ and Ψ' agree to the r-th order at the point p". Since this will obviously be a purely local property, we can restrict ourselves to the case where E is a Cartesian product $M \times E(p)$ of M with the manifold E(p). As we have seen, a "cross-section" then can be associated with a map $\alpha\colon M \to E(p)$, as follows:

$$\Psi(p') = (p', \alpha(p')) \tag{1.1}$$

for all $p' \in M'$.

Now, we can say that two maps $\alpha, \alpha'\colon M \to E(p)$ "agree to the r-th order at p" if

a) $\alpha(p) = \alpha'(p)$

b) With respect to local coordinizations of M and E(p), all partial derivatives of α and α' agree up to the r-th order. (1.2)

Given Ψ, $\Psi' \in \Gamma(E)$, let us construct α, α' as maps: $M \to E(p)$, using (1.1), and say that Ψ *and* Ψ *agree up to the r-th order at* p if α and α' satisfy (1.2). Now, one must verify (exercise) that this notion is independent of the product structure chosen for the fiber space. Once having done this, the reader will readily convince himself that "Ψ and Ψ' agreeing to the r-th order at p" defines an equivalence relation on $\Gamma(E)$. The quotient of $\Gamma(E)$ by this equivalence relation is defined as the space of *r-th order jets of cross-sections at* p, denoted by: $J^r(E)(p)$.

We next want to make the disjoint union of the $J^r(E)(p)$, as p runs over M, into some sort of space. This can be done most conveniently as follows:

Consider the space $\Gamma(E) \times M$. Introduce an equivalence relation as follows:

(Ψ, p) is equivalent to (Ψ', p')

if:

a) $p = p'$

b) Ψ and Ψ' agree to the r-th order at p. (1.3)

The quotient of $\Gamma(E) \times M$ by this equivalence

relation is denoted by: $J^r(E)$, and called the *space of r-th order jets associated with the fiber space* (E, M, π).

The simple properties of $J^r(E)$ are immediately deduced from this definition. First, notice that the projection map $(\Psi, p) \to p$ of $\Gamma(E) \times M \to M$ "passes to the quotient" to define a "projection map":

$$\pi^r: \ J^r(E) \to M.$$

The fiber of π^r over a point $p \in M$ is identifiable with $J^r(E)(p)$, as defined above. (This is implicit in our notation, of course.)

If $r' < r$, then there is also a projection-type map: $J^r(E) \to J^{r'}(E)$, since if $\Psi, \Psi' \in \Gamma(E)$ agree to the r-th order at p, they agree to the r'-th order.

Now, $J^r(E)$ may be considered as a fiber space over M, with π^r the projection map. Let $\Gamma(J^r(E))$ denote its space of cross-sections. Then, there is a map j^r: $\Gamma(E) \to \Gamma(J^r(E))$ defined as follows:

$j^r(\Psi)(p)$ = the equivalence class to which

$(\Psi, p) \in \Gamma(E) \times M$ belongs.

$j^r(\Psi)$ is called the *r-jet* of the cross-section Ψ.

After this brief outline of the generalized formalism we turn to the presentation of certain general features of the calculus of variations. As we shall see, the "jet-bundle" language is very well suited to this task. (Although, it is not necessarily the ideal formalism with which to do explicit calculations.)

2. LAGRANGIANS AND FIBER SPACES

Let π: $E \to M$ be a fiber space, and let r be an integer. An *r-th order Lagrangian* for the fiber space is a real-valued function on $J^r(E)$, which we will usually denote by L.

Suppose that M has a fixed volume element, denoted by dp. Thus, if $p \to f(p)$ is a function on M,

$$\int_M f(p)dp$$

denotes its integral over M.

Given a Lagrangian L, let $\underset{\sim}{L}$ denote the following real-valued function on $\Gamma(E)$:

$$\underset{\sim}{L}(\Psi) = \int_M L(j^r(\Psi)(p))dp. \qquad (2.1)$$

The calculus of variations studies the properties of the function $\underset{\sim}{L}$, on curves in $\Gamma(E)$. The typical problem is to compute the "first variation"

$$\frac{d}{d\lambda} \underset{\sim}{L}(\Psi^\lambda)/_{t=0}, \qquad (2.2)$$

where $\lambda \to \Psi^\lambda$ is a one-parameter family of elements of $\Gamma(E)$. Working out this derivative gives the "Euler-Lagrange differential operator", which maps cross-sections of E into cross-sections of another bundle.

Now, it is difficult to describe the Euler-Lagrange operator in complete generality at this stage, without using coordinates. (We will see one general coordinate free method involving "differential forms" at a later point.) Then, we will proceed to the calculation in the traditional way, using coordinates. In the next sections we will consider various special (but important) cases that can be treated in a coordinate-free way, quite readily.

The special assumptions we will make are:

a) $r = 1$.

b) E is the product: $M \times R^n$.

c) M can be coordinatized by Euclidean coordinates, denoted by $x = (x_\mu)$.

For the sake of applications to quantum field theory, it suffices to deal with the case: $\dim M = 4$, so that the indices (μ, ν) run over the following ranges:

$$0 \leq \mu, \nu \leq 3.$$

Denote coordinates on the fiber of E by $\phi = (\phi_a)$,

$$1 \leq a, b \leq n.$$

Thus, an element $\Psi \in \Gamma(E)$ is determined by functions: $\phi(x) = (\phi_a(x_\mu))$.

We can determine a coordinization of $J^1(E)$ as follows: Define the functions $\phi_{a\mu}$ on $J^1(E)$ as follows:

$\phi_{a\mu}$ assigns to $\Psi \in \Gamma(E)$, which is represented by function $(\phi_a(x_\mu))$, the partial derivatives $\partial_\mu \phi_a(x)$.

(∂_μ denotes $\frac{\partial}{\partial x_\mu}$.)

Thus, $(x_\mu, \phi_a, \phi_{a\mu})$ determine a coordinization of $J^1(E)$ (Exercise). A real-valued function L on $J^1(E)$-what we call a "Lagrangian" - is then represented by a function $L(x, \phi, \phi_{a\mu})$ of the indicated variables.

Suppose also that the volume element of dp for M is the Euclidean volume element: $dx = d^4x$. Then, we have:

$$L(\Psi) = \int L(x, \phi(x), \partial\phi(x))dx \tag{2.3}$$

The left hand side of (2.3) is given in fully "intrinsic" or "covariant" terms by the right hand side of (2.1). The right hand side of (2.3) is the expression used in the "classical" books on the calculus of variations. However, notice that identifying this "classical" expression within the jet-bundle formalism gives us a precisely defined meaning to the "covariance" properties of the expressions used in the classical calculus of variations.

With these conventions, let us proceed to calculate (2.2). Suppose that

$$\lambda \to \Psi^{\lambda} = ({\phi_a}^{\lambda}(x))$$

is a one-parameter family of cross-sections, which reduces to $\Psi = (\phi_a(x))$ at $\lambda = 0$.

Set:

$$\delta_a(x) = \frac{\partial}{\partial\lambda} {\phi_a}^{\lambda}(x)/_{\lambda=0} \tag{2.4}$$

Suppose that $\delta_a(x)$ vanishes if x is sufficiently large. Introduce the following notations:

$$L_a = \frac{\partial L}{\partial \phi_a}$$
$$L_{a\mu} = \frac{\partial L}{\partial \phi_{a\mu}} \, . \tag{2.5}$$

Then, combining (2.2)-(2.5) gives:

$$\frac{d}{d\lambda} L(\Psi^{\lambda})/_{\lambda=0}$$

$$= \frac{d}{d\lambda} \int L(x, \phi^\lambda(x), \partial\phi^\lambda(x))dx /_{\lambda=0}$$

$$= \int [L_a(x, \phi(x), \partial\phi(x))\delta_a(x)]dx$$

$$+ L_{a\mu}(x, \phi(x), \partial\phi(x))\partial_\mu \delta_a(x)]dx$$

= , after integrating by parts and taking into account that the $\delta_a(x)$ vanish if x goes to infinity,

$$\int [L_a(x), \phi(x), \partial\phi(x)) - \partial_\mu(L_{a\mu}(x, \phi(x), \partial\phi(x)))]\delta_a(x)dx \tag{2.6}$$

This motivates the introduction of the *Euler-Lagrange differential operator:*

$$\phi(x) \to [L_a(x, \phi(x), \partial\phi(x) - \partial_\mu(L_{a\mu}(x, \phi(x), \partial\phi(x)))] \tag{2.7}$$

It maps cross-sections of the fiber space (E, M, π) into cross-sections of some other vector bundle. What this bundle is, and whether the operator defined via (2.7) is independent of the special choices made, is unclear at the moment. However, we can say that, if the right hand side of (2.7) vanishes, then:

$$\frac{d}{d\lambda} \underset{\sim}{L}(\Psi^{\lambda})\Big/_{\lambda=0} = 0 \qquad (2.8)$$

for any curve $\lambda \to \Psi^{\lambda}$ in $\Gamma(E)$ beginning at Ψ for $\lambda = 0$, which has suitable vanishing properties "at infinity" for M. Thus, we may define the *extremals* or the cross-sections $\Psi \in \Gamma(E)$ whose representative functions $(\phi_a(x))$ satisfy the differential equations obtained by setting equal to zero the right hand side of (2.7) - the so-called "Euler-Lagrange equations".

We shall now turn to the study of vector bundles.

3. EULER-LAGRANGE OPERATORS ON VECTOR BUNDLES AND ASSOCIATED DIFFERENTIAL OPERATORS

First, we must define the concept of "adjoint differential operator".

Suppose now that π: E → M, is a vector[1]

---

[1]For simplicity, we will suppose that all vector spaces are defined as real vector spaces.

bundle, and that the following sort of structures are given:

a) A volume element dp on M, enabling one to define the integral $\int_M f(p)dp$ of a real-valued function on M

b) A real-valued bilinear form $(v_1, v_2) \to \beta(v_1, v_2)$ on each fiber E(p) of E.

Let $\Gamma(E)$ denote the vector space of cross-sections of E, while $\Gamma_o(E)$ denotes the subspace of those with compact support, i.e. those Ψ: $M \to E$ that vanish outside of some compact subset of M.

Given $\Psi_1 \in \Gamma_o(E)$, $\Psi \in \Gamma(E)$, define their "inner product" as:

$$\underset{\sim}{\beta}(\Psi_1, \Psi_2) = \int_M \beta(\Psi_1(p), \Psi_2(p))dp \qquad (3.1)$$

Suppose now that π': $E' \to M$ is another vector bundle, with a form β' defined on its fibers, and an "inner product" $\underset{\sim}{\beta}'(\Psi_1', \Psi_2')$ defined similarly to (3.1). Suppose also that D: $\Gamma(E) \to \Gamma(E')$ is a linear differential operator.

<u>DEFINITION</u>. A linear differential operator D^*: $\Gamma(E') \to \Gamma(E)$ is said to be the *adjoint* of D (relative to the inner products β, β') if:

$$\underset{\sim}{\beta}'(\Psi_1', D(\Psi_2)) = \underset{\sim}{\beta}(D^*(\Psi_1'), \Psi_2)$$
$$\text{for } \Psi_1' \in \Gamma(E), \Psi_2 \in \Gamma_o(E) \qquad (3.2)$$

By making everything explicit in terms of local coordinates and product structures for the bundle, one sees readily that such an adjoint operator exists, and is of the same order as D. Further, the following rules hold:

$$(D_1 + D_2)^* = D_1^* + D_2^*$$
$$(D_1 D_2)^* = D_2^* D_1^* \qquad (3.3)$$
$$(D^*)^* = D.$$

Using these adjoint operators, we can readily calculate the Euler-Lagrange operators, in case the Lagrangian is defined by differential operators. Consider a Lagrangian of the following form:

$$\underset{\sim}{L}(\Psi) = \underset{\sim}{\beta}'(D_1\Psi, D_2\Psi) \qquad (3.4)$$

where $\Psi \in \Gamma_o(E)$, and D_1, D_2 are differential operators: $\Gamma(E) \to \Gamma(E')$, and $\underset{\sim}{\beta}'$, $\underset{\sim}{\beta}$ are *symmetric* inner products. Suppose that $\lambda \to \Psi^\lambda$ is a curve in $\Gamma_o(E)$, with:

$$\Psi^o = \Psi; \ \delta = \frac{d}{d\lambda} \Psi^\lambda /_{\lambda=0}.$$

Then,

$$\frac{d}{d\lambda} \underset{\sim}{L}(\Psi^\lambda)/_{\lambda=0} = \frac{d}{d\lambda} \beta'(D_1\Psi^\lambda, D_2\Psi^\lambda)/_{\lambda=0}$$

$$= \beta'(D_1\delta, D_2\Psi) + \beta'(D_1\Psi, D_2\delta)$$

$$= \beta'(\delta, D_1^* D_2\Psi) + \beta'(D_2^* D_1\Psi, \delta)$$

$$= \beta'(D_1^* D_2\Psi + D_2^* D_1\Psi, \delta).$$

Now, set:

$$D_L = D_1^* D_2 + D_2^* D_1 \qquad (3.5)$$

It is the "Euler-Lagrange operator" associated with (3.4). That it can be expressed so simply is one of the virtues of the differential operator formalism!

Of course, if we are given an Lagrangian that is the sum $\underset{\sim}{L} = \underset{\sim}{L}_1 + \underset{\sim}{L}_2 + \ldots$ where $\underset{\sim}{L}_1, \underset{\sim}{L}_2, \ldots$ are each of type (3.4), then the Euler-Lagrange operator D_L associated with $\underset{\sim}{L}$ may be considered as the sum:

$$D_L = D_{L_1} + D_{L_2} + \ldots \qquad (3.6)$$

4. SYMMETRY GROUPS AND CONSERVED CURRENTS FOR GENERAL LAGRANGIANS

Let us revert to the explicit, coordinate notations of Section 2. Introduce again variables $\phi = (\phi_a)$, $x = (x_\mu)$, $(\phi_{a\mu}) = \dot{\phi}$

$$1 \leq a, b \leq n; \; 0 \leq \mu, \nu \leq 3.$$

Consider a Lagrangian function $L(\phi, \partial\phi)$, and the associated Euler-Lagrange differential operator:

$$\partial_\mu(L_{a\mu}) - L_a.$$

Suppose that $(\phi_a{}^\lambda(x))$, $0 \leq \lambda \leq 1$, is a one-parameter family of "extremals", i.e. solutions of the Euler-Lagrange differential equations. Thus, we have:

$$\partial_\mu(L_{a\mu}(\phi^\lambda, \partial\phi^\lambda)) - L_a(\phi^\lambda, \partial\phi^\lambda) = 0 \qquad (4.1)$$

for all λ.

Suppose also that:

$$\frac{d}{d\lambda} L(\phi^\lambda, \partial\phi^\lambda)dx = 0, \qquad (4.2)$$

i.e. the Lagrangian takes the same value for all λ.

To work out the consequence of these natural[1] assumptions, let us work out the left hand side of (4.2), after setting:

$$\phi_a(x) = \phi_a^{\,o}(x)$$

$$\gamma_a(x) = \frac{\partial}{\partial\lambda} \phi_a^{\,\lambda}(x)/_{\lambda=0}.$$

Then, we have:

[1]Notice, for example, that they would be satisfied if ϕ^λ results from applying to ϕ^o a one-parameter group of symmetries of the Lagrangian.

$$0 = L_a(\phi, \partial\phi)\gamma_a(x) + L_{a\mu}(\phi, \partial\phi)\partial_\mu\gamma_a(x)$$

$$= \text{, using (4.1) at } \lambda = 0,$$

$$\partial_\mu(L_{a\mu}(\phi, \partial\phi))\gamma_a(x) + L_{a\mu}\partial_\mu\gamma_a,$$

or:

$$\partial_\mu(L_{a\mu}(\phi(x), \partial\phi(x))\gamma_a(x)) = 0 \tag{4.3}$$

Set:

$$V_\mu(x) = L_{a\mu}(\phi(x), \partial\phi(x))\gamma_a(x) \tag{4.4}$$

Then, (4.3) expresses the part that the collection of functions $V_\mu(x)$ forms what the physicists call a "conserved current". In differential geometric language, $V_\mu(x)$ can be identified with a 3-differential form ω:

$$\omega = V_o dx_1 \wedge dx_2 \wedge dx_3 - V_1 dx_o \wedge dx_2 \wedge dx_3$$
$$+ V_2 dx_1 \wedge dx_2 \wedge dx_3 - V_3 dx_o \wedge dx_1 \wedge dx_3 \tag{4.5}$$

(4.3) is then equivalent to the condition that that

the 3-form ω be closed, i.e.

$$d\omega = 0, \tag{4.6}$$

where $d\omega$ is the exterior derivative of the differential form ω. In turn, this leads to the following idea:

Consider M as R^4, the "base" of the fiber space whose cross-sections we are identifying with the functions $\phi(x) = (\phi_a(x))$. Let N be a submanifold of M of dimension 3. Set:

$$\underset{\sim}{f}(\phi, N) = \int_N \omega \tag{4.7}$$

the integral of ω over N. (See "Differential geometry and the calculus of variations [2]" for the background here.) Then, $\underset{\sim}{f}$ defined by (4.7) is a real-valued function of the extremal ϕ and the submanifold N. However, if N' is another submanifold such that N and N' "cobound", i.e. are the boundaries of a four dimensional region, then - using Stokes' theorem and (4.6) -

$$\underset{\sim}{f}(\phi, N) = \underset{\sim}{f}(\phi, N') \tag{4.8}$$

This property is the geometric version of what the physicists call a "conserved current".

For example, if we choose N as the submanifolds: $x_o = t$: , with t a constant, then:

$$\underset{\sim}{f}(\phi, N) = \int V_o(t, x_1, x_2, x_3)dx_1dx_2dx_3 \quad (4.9)$$

The "conserved" condition then expresses the fact that the right hand side of (4.9) is independent of t. Physically, the right hand side of (4.9) is often called the "charge" generated by the current (4.3) then expresses the fact that "charge" is conserved.

The $\gamma_a(x)$ can often be made more explicit in terms of groups of transformations. Let E be the space of variables (x, ϕ), i.e. E is the "total space" of the fiber space. Suppose that $\lambda \to g(\lambda)$ is a one-parameter group of bundle isomorphisms on E, whose infinitesimal generator vector field, X, is of the form:

$$X = A_\mu(x) \frac{\partial}{\partial x_\mu} + A_a(\phi, x) \frac{\partial}{\partial \phi_a} , \quad (4.10)$$

where A_μ, A_a are functions of the indicated variables.

Then,

$$\gamma_a(x) = A_\mu(x)\partial_\mu\phi_a(x) + A_a(\phi(x), x),$$

so that:

$$V_\mu(x) = (L_{a\mu}A_\nu\phi_{a\nu} + A_aL_{a\mu})(j^1(\Psi)(x)) \quad (4.11)$$

This notation can be explained as follows: Regard (x, ϕ_a) as coordinates of E, a fiber space over R^4. Regard $(x, \phi_a, \phi_{a\mu})$ as coordinates of $J^1(E)$, the space of 1-jets of cross-sections.

L, L_a, $L_{a\mu}$, A_a, A_μ then are functions on $J^1(E)$. If Ψ: $R^4 \to E$ is a cross-section (represented by functions $x \to (\phi_a(x))$, then $j^1(\Psi)$ - its "1-jet" - is a cross-section of $J^1(E)$, represented by the functions $(\phi_a(x), \phi_{a\mu}(x) = \partial_\mu\phi_a(x))$.

There is also another important geometric and physical interpretation of those calculations. Consider the submanifold N held fixed, and regard $\phi \to \underset{\sim}{f}(\phi, N)$ as a real-valued function on the space of extremals of the Lagrangian L, i.e. the solution of the Euler-Lagrange differential equations. Using a "Poisson-bracket", the space of these functions can be made into a Lie algebra.

The "quantum field theory" generated by the Lagrangian L should then be a representation of a certain Lie algebra (under Poisson bracket) of these functions by operators on a "Hilbert space". We shall study this point of view at a later point. (See also the material in "Lie algebras and quantum mechanics".)

CHAPTER III

CONNECTIONS IN VECTOR BUNDLES AND YANG-MILLS FIELDS

It is well-known that the theory of what the physicists call "Yang-Mills fields" is almost identical to the theory of "linear connections in vector bundles". (See, for example, "Fourier analysis on groups and partial wave analysis", [3].) Also, what the physicists call a "minimal electromagnetic coupling" also essentially involves connection ideas. Connections provide an important and illustrative example to which the general theory of fiber spaces and vector bundles applies very neatly. With these considerations as motivation, we turn to their study.

1. LINEAR CONNECTIONS IN VECTOR BUNDLES

For this chapter, π: $E \to M$ will denote a vector bundle. As usual, $\Gamma(E)$ denotes its space of cross-sections. The "differential geometric" point of view - regarding $\Gamma(E)$ as a module over the algebra $F(M)$ of real-valued, C^∞ functions on M - will dominate the work in this chapter. (See "Differential geometry and the calculus of variations" [2].) V(M) denotes the Lie algebra of vector fields on M, i.e. the space of first order, homogeneous differential operators: $F(M) \to F(M)$.

DEFINITION. *A (Linear) connection* for the vector bundle (E, M, π) is defined as a real-bilinear map $V(M) \times \Gamma(E) \to \Gamma(E)$, satisfying condition (1.1) described below. A typical connection is denoted by ∇. Given $X \in V(M)$, $\Psi \in \Gamma(E)$, the image under ∇ of (X, Ψ) is denoted by: $\nabla_X\Psi$: This is read as "the covariant derivative of Ψ by X", and is again a cross-section of E.

The characteristic conditions for a connection are:

$$\text{a)}\quad \nabla_{fX}\Psi = f\nabla_X\Psi$$
$$\text{b)}\quad \nabla_X(f\Psi) = X(f)\Psi + f\nabla_X(\Psi) \qquad (1.1)$$

for $f \in F(M)$, $X \in V(M)$, $\Psi \in \Gamma(E)$.

(One of the great virtues of the "F(M)-module" point of view is that the idea of a connection can be presented so simply!)

Let us now show how conditions (1.1) can be used to describe connections more explicitly in terms of components. (In classical tensor analysis - and for certain types of connections - those components would be called "Christoffel symbols".) The "components" can be most simply described by choosing an F(M)-module basis (Ψ_a), for $\Gamma(E)$, with the following range of indices and summation connections:

$$1 \leq a, b \leq n.$$

By the definition of "basis" (see Chapter 1), for each $X \in V(M)$ there are functions $\gamma_{ab}(X) \in F(M)$ such that:

$$\nabla_X(\Psi_a) = \gamma_{ab}(X)\Psi_b \qquad (1.2)$$

Choose an F(M)-basis (X_i), $1 \leq i, j \leq m$, for V(M). Set:

$$\gamma_{abi} = \gamma_{ab}(X_i) \tag{1.3}$$

Then, the rules (1.1) guarantee that the connection ∇ is completely determined by the functions (γ_{abi}). Thus, they may be called the *components* of ∇ with respect to the "moving frames" (Ψ_a, X_i). Conversely, the steps are reversible: Any such set of functions determines a connection. Notice, however, that the "components" γ_{abi} of ∇ do not have a linear, tensorial form when these module-bases are changed. (*Exercise:* Determine explicitly the change in γ_{abi} when "new" module bases (Ψ_a', X_i') are chosen. However, we will never need to work with this "classical-tensor analysis" point of view.)

Let $\underset{\sim}{C}(E)$ denote the set of these linear connections. The "non-tensorial" nature of connections is related to the fact that $\underset{\sim}{C}(E)$ is not in a "natural" way an F(M)-module. Instead, it has a different sort of algebraic structure, which we will examine further on.

As we have just seen, a connection does not form a "tensor". However, one can associate various tensors with a connection. The simplest of these tensors is the "curvature tensor".

<u>DEFINITION</u>. Let $\nabla \varepsilon \underset{\sim}{C}(E)$ be a linear connection for a vector bundle (E, M, π). For $X, Y \varepsilon V(M)$, $\Psi \varepsilon \Gamma(E)$, set:

$$R_\nabla(X, Y)(\Psi) = \nabla_X \nabla_Y(\Psi) - \nabla_Y \nabla_X(\Psi) - \nabla_{[X, Y]}(\Psi) \tag{1.4}$$

Thus, for each $X, Y \varepsilon V(M)$, $R_\nabla(X, Y)$ is an R-linear mapping: $\Gamma(E) \to \Gamma(E)$. We will often abbreviate this to: R_∇, or $R_\nabla(\ ,\ )(\ )$. R_∇ is called the *curvature tensor* associated with the connection. Thus, R_∇ is an R-trilinear map:
$V(M) \times V(M) \times \Gamma(E) \to \Gamma(E)$.

Now, a-priori, one would not expect from (1.1) that R_∇ has "better" tensorial properties than ∇ itself. However, it does. In fact,

$$\begin{aligned} &R_\nabla(fX, Y)(\Psi) = fR_\nabla(X, Y, \Psi) \\ &= R_\nabla(X, fY)(\Psi) = R_\nabla(X, Y)(f\Psi) \end{aligned} \tag{1.5}$$

for $X, Y \in V(M)$, $f \in F(M)$, $\Psi \in \Gamma(E)$.

The verification of (1.5) is a routine calculation - based on the rules (1.1) - and is left to the reader. (This calculation is done for the special case of an "affine connection - that is, a connection in the tangent bundle T(M) of the manifold M - in "Differential geometry and the calculus of variations". The general case is similar.)

The rules (1.5) guarantee a "tensorial" transformation law for the components of R. Precisely, if (Ψ_a, X_i) are F(M)-module bases for $\Gamma(E) \times V(M)$, and if:

$$R_\nabla(X_i, X_j)(\Psi_a) = R_{ijab}\Psi_b, \tag{1.6}$$

then the components (R_{ijab}) change "linearly", i.e. tensorially, under change of bases.

The curvature tensor R_∇ itself may then be considered as a cross-section of a vector bundle (E', M, π') whose base space is M. (The fiber of E' above a point $p \in M$ is the space of all trilinear maps: $M_p \times M_p \times E(p)$, where M_p denotes the tangent space to M at p.) Then, the map $\nabla \to R_\nabla$ of

$\underset{\sim}{C}(E) \to \Gamma(E')$ may be considered as a "differential operator" (although a "non-linear" one).

We now ask: What is the meaning of the condition that the curvature tensor be zero? As a first remark, notice from (1.1) directly that the following condition *implies* that $R_\nabla = 0$:

> There exists an F(M)-module bases (Ψ_a) of $\Gamma(E)$ such that

$$\nabla_X \Psi_a = 0 \quad \text{for all} \quad X \in V(M). \tag{1.7}$$

Conversely, suppose that $R_\nabla = 0$. Let (Ψ'_a) be an arbitrary F(M)-module bases for $\Gamma(E)$. Let us look for a basis (Ψ_a) satisfying (1.7), of the form:

$$\Psi_a = f_{ab}\Psi'_b \tag{1.8}$$

where (f_{ab}) is a matrix of functions on M. Suppose that:

$$\nabla_X(\Psi'_a) = \gamma_{ab}(X)\Psi'_b \tag{1.9}$$

Using (1.1), we have:

$$\nabla_X(\Psi_a) = X(f_{ab})\Psi'_b + f_{ab}\gamma_{bc}(X)\Psi'_c \qquad (1.10)$$

Thus, we have:

Condition (1.5) is satisfied if and only if the (f_{ab}) satisfy the following system of differential equations.

$$X(f_{ab}) + f_{ac}\gamma_{cb}(X) = 0 \qquad (1.11)$$

for all $X \varepsilon V(M)$.

Now, we can prove:

THEOREM 1.1. If M is simply connected[1], if p_o is an arbitrary point of M, and if $(f_{ab}{}^o)$ is an arbitrary matrix of real numbers with non-zero determinant, then there is a unique solution of the equation (1.11), for the (f_{ab}), reducing to $(f_{ab}{}^o)$ at $p = p_o$.

---

[1]A space is *simply connected* if each closed continuous curve in the space can be shrunk continuously to a point.

Proof. We shall construct (f_{ab}) by finding systems of linear, ordinary differential equation it must satisfy if it did exist.

Suppose that $t \to \sigma(t)$, $0 \le t \le 1$, is a curve in M, beginning at p_o for $t = 0$. Suppose $\sigma(t) = p$, another fixed (for the moment) point of M. Let $\sigma'(t) \in M_{\sigma(t)}$ denote the tangent vector to the curve at time t. Choose X as a vector field on M such that:

$$\sigma'(t) = X(\sigma(t)) \quad \text{for} \quad 0 \le t \le 1. \tag{1.12}$$

(In differential geometric language, σ is an *integral curve* for X.) Now, by the definition of "tangent vector to a curve", relation (1.1) means that, for each $f \in F(M)$,

$$\frac{d}{dt} f(\sigma(t)) = X(f)(\sigma(t)) \tag{1.13}$$

Combining (1.9) and (1.13) gives:

$$\frac{d}{dt} f_{ab}(\sigma(t)) + f_{ac}(\sigma(t))\gamma_{cb}(\sigma'(t)) = 0 \tag{1.14}$$

Notice that the $\gamma_{cb}(\sigma'(t))$ are functions of t that only depend on the curve σ (and the connection, of course). Thus, (1.14) is a system of linear ordiary differential equations for the functions $f_{ab}(\sigma(t))$. From the theory of such systems, we know that there is a unique solution, reducing to f_{ab}^{o} at $t = 0$. Suppose that we now define

$$f_{ab}(p) = f_{ab}(\sigma(1)), \tag{1.15}$$

where the right hand side of (1.15) is the value at $t = 1$ of the solution of (1.14).

A-priori, $(f_{ab}(p))$ depends on the curve joining p_o to p. Here is where the condition that: a) M be simply connected; and: b) $R_\nabla = 0$: enters. Suppose that $\sigma_1(t)$ is another curve joining p_o to p. By the "simply connected" condition, σ can be deformed into σ_1. This means that there exists a "homotopy" $(s, t) \to \delta(s, t)$, $0 \le s, t \le 1$, in M such that:

$$\delta(0, t) = \sigma(t); \ \delta(1, t) = \sigma_1(t)$$

$$\delta(s, 0) = p_o, \ \delta(s, 1) = p.$$

For s held constant let $(f_{ab}^{s}(t))$ be the solution

of (1.14), derived from the curve σ^s: $t \to \delta(s, t)$. To show that $f_{ab}(p)$ is indeed independent of the curve chosen to join p_o to p we must show that:

$$\frac{\partial}{\partial s}(f_{ab}{}^s(1)) = 0 \tag{1.16}$$

Now, we have, from its definition, that:

$$\frac{\partial}{\partial t} f_{ab}(\sigma^s(t)) + f_{ac}(\sigma^s(t))\gamma_{cb}(\sigma^{s\,\prime}(t)) = 0.$$

This relation must now be differentiated with respect to s. Substituting into the resulting relation the condition: $R_\nabla = 0$: will give a system of linear ordinary differential equations for the functions $t \to \frac{d}{ds} f_{ab}{}^s(t)$, from which one derives (1.16). Similar calculations are presented in "Differential geometry and the calculus of variations", hence will not be repeated here.

Another point of view uses the "Frobenius complete integrability theorem" to solve (1.11) for the (f_{ab}). Let G be the group $GL(n, R)$ of $n \times n$ real, invertible matrices.

Set: $N = M \times G$: Regard $(f_{ab}(p))$ as an element of

G. Then, a matrix of functions (f_{ab}) defines a map ϕ: $M \to N$ by assigning to p the point $(p, (f_{ab}(p)))$ of $N = M \times G$. One shows that there is a Pfaffian system of differential forms on N, such that ϕ is an integral manifold of it if and only if f_{ab} satisfy (1.11). One further shows that the conditions of "complete integrability" of this Pfaffian system are precisely equivalent to: $R_\nabla = 0$: Theorem 1.1 would then follow (at least locally) from the local version of the Frobenius theorem. One also proves that a "leaf" of the resulting "foliation", when projected down to M, would be a covering map onto M: If M were simply connected, this projection of a leaf onto M would be a diffeomorphism, and the inverse map, when projected into G, would be the desired solution of (1.13).

2. THE FIBER SPACE WHOSE CROSS-SECTIONS ARE CONNECTIONS

Let (E, M, π) continue to be a fixed vector bundle, with $\underset{\sim}{C}(E)$ denoting its space of cross-sections. So far, we have been considering a fixed connection $\nabla \in \underset{\sim}{C}(E)$. The "Yang-Mills idea" - which is in turn closely related to the ideas interrelating geometry and physics arising out of general relativity theory - is to regard connections as fields also. In particular, we will want to consider "Lagrangians" for the space of connections. The general ideas of Chapter 2 will provide us with a way of doing this providing we can interpret $\underset{\sim}{C}(E)$ as the space of cross-sections of some fiber space. We will now show how to do this.

Consider two connections $\nabla, \nabla' \in \underset{\sim}{C}(E)$. Now, the sum $\nabla + \nabla'$ is not a connection since (1.1) is not longer satisfied. However, let us define:

$$\tau = \nabla - \nabla' \tag{2.1}$$

i.e. τ is the "difference". It, too, is no longer a connection. However, notice that - even better - it is a "tensor field".

First, formula (2.1) indicates that τ is an R-bilinear map: $V(M) \times \Gamma(E) \to \Gamma(E)$

$$\tau(X, \Psi) = \nabla_X \Psi - \nabla'_X \Psi \tag{2.2}$$

for $X \in V(M)$, $\Psi \in \Gamma(E)$

Notice that the "bad" - i.e. non-tensorial-term on the right hand side of (1.1) cancels out, and τ satisfies:

$$\tau(fX, \Psi) = f\tau(X, \Psi) = \tau(X, f\Psi) \quad \text{for} \quad X \in V(M),\ \Psi \in \Gamma(E),\ f \in F(M) \tag{2.3}$$

(2.3) means that τ is an F(M)-bilinear mapping. Thus, τ has a "value" at each point $p \in M$, denoted by τ_p, which is an R-bilinear map: $M_p \times E(p) \to E(p)$ such that:

$$\tau_p(X(p), \Psi(p)) = \tau(X, \Psi)(p) \quad \text{for} \quad p \in M,\ X \in V(M),\ \Psi \in \Gamma(E). \tag{2.4}$$

Thus, τ is interpretable as a cross-section of a vector bundle E" over M; the fiber of E" over a point $p \in M$ is the space of all bilinear maps: $N_p \times E(p) \to E(p)$.

<u>DEFINITION</u>. Two affine connections ∇, $\nabla' \in C(E)$ *agree at a point* $p \in M$ *if:*

$$\tau_p = 0 \tag{2.5}$$

where τ_p is the "value" at p of the tensor field defined by (2.1).

Now, we can define a fiber space C(E) over M, whose cross-sections can be identified with the connections of (E, M, π), in the following way:

> Set up an equivalence relation in $\underset{\sim}{C}(E) \times M$: (∇, p) is equivalent to (∇', p') if and only if:
>
> a) $p = p'$
>
> b) ∇ and ∇' agree at p.

C(E) is defined as the quotient of $\underset{\sim}{C}(E) \times M$ by this equivalence relations.

This sets us up to apply the general theory. C(E) is a fiber space over M; we can define $J^1(C(E))$ as usual, define Lagrangians as functions on $J^1(C(E))$, calculate the Euler-Lagrange operators associated with these Lagrangians, etc. This will give us "dynamical equations", whose "solutions" will be connections, thus promoting "connections" to the status of physical quantities. This, in essence, is the geometric version of the "Yang-Mills idea".

3. CURVATURE AS A FUNCTION ON THE JET-BUNDLE

Continue the notations of Section 2. Thus, $\underset{\sim}{C}(E)$ is identified with the space of cross-sections, $\Gamma(C(E))$, of a fiber space $C(E)$ over M. Recall that $J^1(C(E))$ is defined in the following way:

> Introduce an equivalence relation into $\underset{\sim}{C}(E) \times M$ as follows: (∇, p) is equivalent to (∇', p') if and only if:
>
> a) $p = p'$
>
> b) $\nabla - \nabla' = \tau$ is zero at p
>
> c) τ vanishes to the first order at p, in the sense that the first derivatives of its components with respect to moving frames vanish at the point p.

Then, of course, $J^1(C(E))$ is defined as the quotient of $\underset{\sim}{C}(E) \times M$ by this equivalence relation.

Given $\nabla \in \underset{\sim}{C}(E) = \Gamma(C(E))$, construct R_∇, its curvature tensor. Let (E'', M, π'') be the vector bundle over M for which R_∇ is a cross-section. One sees readily that, if ∇ and ∇' "agree to the first order", in the sense of b) and c), that

$R_{\nabla} = R_{\nabla'}$ at p.

Thus, R_{∇} "passes to the quotient", to define a fiber-preserving map: $J^1(C(E)) \to E''$. In turn, this enables us to construct Lagrangian on C(E) by taking arbitrary[1] functions on E", and pulling them back to $J^1(C(E))$ via this map. We will show how to construct the corresponding Euler-Lagrange operators in the next sections.

4. DEFORMATIONS OF CURVATURE AND LINEAR CONNECTIONS

Let (E, M, π) continue to be a vector bundle, with $\nabla \in \underset{\sim}{C}(E)$ a linear connection. Consider a one-parameter family $\lambda \to \nabla^{\lambda}$ of such connections, reducing to ∇ at $\lambda = 0$.

Set:

$$\delta = \frac{\partial}{\partial\lambda} \nabla^{\lambda}/_{\lambda=0}, \tag{4.1}$$

---

[1]For physical applications, one will want to choose these functions so that they will transform properly under certain transformations groups.

i.e. δ is a map $V(M) \times \Gamma(E) \to \Gamma(E)$, defined by:

$$\delta(X, \Psi) = \frac{\partial}{\partial\lambda} \nabla_X^\lambda(\Psi)/_{\lambda=0}. \qquad (4.2)$$

Again, one sees that δ is a "tensor field", i.e. a cross-section of a vector bundle (E', M, π') over M.

Let R_{∇^λ} be the curvature tensor of ∇^λ, and set:

$$\rho = \frac{d}{d\lambda} R_{\nabla^\lambda}/_{\lambda=0}. \qquad (4.3)$$

Again, ρ is a cross-section of a vector bundle (E'', M, π''). Now, one sees readily that ρ is determined in terms of δ, and - for fixed ∇ - it is given as the image under ρ of a linear differential operator

$$D_\nabla: \quad \Gamma(E') \to \Gamma(E'') \qquad (4.4)$$

Our aim now is to calculate this operator.

This can readily be done using the explicit formula for the curvature tensor.

$$R_{\nabla^\lambda}(X, Y)(\Psi) = \nabla_X^\lambda \nabla_Y^\lambda \Psi - \nabla_Y^\lambda \nabla_X^\lambda \Psi - \nabla^\lambda_{[X, Y]}\Psi \tag{4.4}$$

Differentiating (4.4) with respect to λ, and setting $\lambda = 0$ gives:

$$\rho(X, Y)(\Psi) = \delta(X, \nabla_Y\Psi) + \nabla_X(\delta(Y, \Psi)) - \delta(Y, \nabla_X\Psi) - \nabla_Y(\delta(X,\Psi)) - \delta([X, Y], \Psi) \tag{4.5}$$

for $X, Y \in V(M)$, $\Psi \in \Gamma(E)$.

Then, we define $D_\nabla\colon\ \Gamma(E') \to \Gamma(E'')$ by the formula:

$(D_\nabla\delta)(X, Y)(\Psi)$ = right hand side of (4.5), for an arbitrary $\delta \in \Gamma(E')$. (4.6)

This explicit formula shows that D is a first order, linear differential operator.

This formula can be simplified in case ∇ - as a linear connection on E - is associated with a linear connection on the tangent bundle to M, i.e. if there is a covariant derivative operation $(X, Y) \to \nabla'_X Y$ for vector bundles on M. Using

$\underset{\sim}{\nabla} \varepsilon\, C(E)$ and $\nabla' \varepsilon\, \underset{\sim}{C}(T(M))$, we can define a "covariant derivative" operator $\nabla''_Y\colon\ \Gamma(E') \to \Gamma(E')$ as follows:

$$\nabla''_Y(\delta)(X, \Psi) = \nabla_Y(\delta(X, \Psi))$$

$$- \delta(\nabla_Y, X, \Psi) - \delta(X, \nabla_Y\Psi) \qquad (4.7)$$

for $X, Y \varepsilon V(M), \quad \Psi \varepsilon \Gamma(E).$

Suppose also that ∇' is "torsion-free", which means that:

$$\nabla'_X Y - \nabla_Y X = [X, Y] \qquad (4.8)$$

for $X, Y \varepsilon V(M)$

Combining (4.5)-(4.8) now gives:

$$(D_\nabla\delta)(X, Y)(\Psi) = \delta(X, \nabla_Y\Psi) + \nabla''_X(\delta)(Y, \Psi)$$

$$+ \delta(\nabla'_X Y, \Psi) + \delta(Y, \nabla_X\Psi) - \delta(Y, \nabla_X\Psi)$$

$$- \nabla''_Y(\delta)(X, \Psi) - \delta(\nabla' X, \Psi) - \delta(X, \nabla_Y\Psi) \qquad (4.9)$$

$$- \delta([X, Y], \Psi)$$

$$= \nabla''_X(\delta)(Y, \Psi) - \nabla''_Y(\delta)(X, \Psi)$$

for $X, Y \varepsilon V(M), \Psi \varepsilon \Gamma(E)$

This is obviously an optimally elegant formula for the differential operator D_∇.

Now, we turn to the question of computing Euler-Lagrange operators.

5. EULER-LAGRANGE OPERATORS ASSOCIATED WITH THE CURVATURE TENSOR

Continue with the notation of the preceding sections. Suppose now that f: $E'' \to R$ is a real-valued function on E". (Recall that E" is the vector bundle over M whose cross-sections are curvature tensors of connections.) Now, E" is a manifold, hence "df", the "differential of f", is a real-valued function on the tangent bundle to E". However, there is a special feature caused by the fact that E" is fibered by vector spaces, since a point of a vector space can be identified with an element of its tangent space also. In fact, if $\lambda \to R^\lambda$ is a curve in the fiber E"(p), with $R^o = R \in E''(p)$,

$$\rho = \frac{d}{d\lambda} \rho^\lambda /_{\lambda=0}$$

there is a real-valued function $(R, \rho) \to df(R, \rho)$

such that:

$$df(R, \rho) = \frac{d}{d\lambda} f(R^{\lambda})/_{\lambda=0} \tag{5.1}$$

Further $df(R, \rho)$ is *linear* in ρ.

Now, suppose that $\lambda \to \nabla^{\lambda}$ is a one-parameter family of linear connections for the vector bundle (E, M, π); that $\delta^{\lambda} = \frac{d}{d\lambda} \nabla^{\lambda}$, and $R_{\nabla^{\lambda}}$, $\rho_{\nabla^{\lambda}} = \frac{d}{d\lambda} R_{\nabla^{\lambda}}$ are defined as in previous sections. Construct a "Lagrangian" function $\underset{\sim}{L}$: $C(E) \to R$ as follows:

$$\underset{\sim}{L}(\nabla) = \int_M f(R_{\nabla}(p))dp. \tag{5.2}$$

Our goal is now to compute:

$$\frac{d}{d\lambda} \underset{\sim}{L}(\nabla^{\lambda})/_{\lambda=0}. \tag{5.3}$$

Now, from (5.2), (5.3) can be written:

$$\int_M \frac{d}{d\lambda} f(R_{\nabla^{\lambda}}(p))dp$$

= , using (5.1),

$$\int_M df(R_{\nabla^\lambda}(p),\ \rho_{\nabla^\lambda}(p))dp/_{\lambda=0}$$
$$= \int_M df(R_{\nabla^\lambda}(p),\ D_\nabla \delta(p))dp. \qquad (5.4)$$

To get the Euler-Lagrange operator explicitly, the operator D_∇ must now be "integrated by parts" to act on R_∇. This will depend on the explicit form for f, of course. We will work it out in an explicit case, namely:

$$f(R) = \beta''(R(p),\ R(p)), \qquad (5.5)$$

where $\beta''(\ ,\ )$ is a symmetric, R-bilinear form on the fibers of the vector bundle. Then,

$$\frac{d}{dt} L(\nabla^\lambda)/_{\lambda=0} = \frac{d}{d\lambda} \int_M \beta''(R_{\nabla^\lambda},\ R_{\nabla^\lambda})dp$$
$$= 2 \int_M \beta''(R_\nabla,\ \rho_\nabla)dp$$
$$= 2 \int_M \beta''(R_\nabla,\ D_\nabla(\delta_\nabla))dp$$
$$= 2 \int_M \beta'(D^*_\nabla R_\nabla,\ \delta_\nabla)dp,$$

where $\beta'(\ ,\ )$ is a symmetric, R-bilinear form on

the fibers of E', and D^*_∇ is the adjoint of the differential operator D_∇ with respect to the forms $\beta''(\ , \ )$ and $\beta'(\ , \ )$, and the volume element form dp. This result is worth stating explicitly.

THEOREM 5.1. With the above notations, the "Euler-Lagrange operator" is the differential operator D^*_∇.

6. THE PRINCIPLE OF MINIMAL INTERACTIONS

Now, we will discuss the way that linear connections are used to introduce "interactions" between Yang-Mills-type fields and more traditional linear fields. (This is the "principle of minimal interaction".)

Let $\pi\colon E \to M$ be a vector bundle. We have already defined $J^1(E)$, the bundle of first order jets, and shown how elegant it is to describe a "Lagrangian" as a real-valued function on $J^1(E)$.

Consider two linear connections $\nabla, \nabla' \in \underset{\sim}{C}(E)$, and a Lagrangian $L \in F(J^1(E))$. The "principle of minimal interaction" assigns to the pair (∇, ∇') and to the Lagrangian L a new Lagrangian $L_{\nabla, \nabla'}$.

The Euler-Lagrange differential operator associated with this Lagrangian physically represents the result of "interacting" the field generated by the Euler-Lagrange operator associated with L and the "field" represented by $\tau = \nabla - \nabla'$.

To see how $L_{\nabla, \nabla'}$ is constructed, we must examine the structure of $J^1(E)$ in more detail. To this end, suppose that a single linear connection $\nabla \in J^1(E)$ is given. Recall that a $\Psi \in \Gamma(E)$ "vanishes to the first order" at p if:

a) $\Psi(p) = 0$

b) All first derivatives of Ψ vanish at $p = 0$.

Now, condition b) can alternately be expressed as follows:

b') $\nabla_X \Psi(p) = 0$ for all $X \in V(M)$.

Then, given $X \in V(M)$, we can define a mapping

$$\nabla_X(p)\colon\ J^1(E)(p) \to E(p) \tag{6.1}$$

by assigning to $j^1(\Psi)(p)$ [1] the element $\nabla_X \Psi(p)$ of $E(p)$.

<u>DEFINITION</u>. For $p \in M$, let $J_\nabla{}^1(E)(p)$ denote the space of vectors in $J^1(E)(p)$ that are annihilated under the maps (6.1) for all $X \in V(M)$, i.e.

$$\nabla_X(p)(j^1(\Psi)) = 0 \quad \text{for all} \quad X \in V(M) \tag{6.2}$$

<u>THEOREM 6.1</u>. $J^1(E)(p)$ is the direct sum of the vector space $J_\nabla{}^1(E)(p)$ and the kernel of the quotient map $J^1(E)(p) \to E(p)$.

Then, as p varies, the union of the $J_\nabla{}^1(E)(p)$ and the $J_h{}^1(E)(p)$ define vector bundles over M, that may be denoted by: $J_\nabla{}^1(E)$ and $J_h{}^1(E)$: We have:

$$J^1(E) = J_\nabla{}^1(E) \oplus J_h{}^1(E), \tag{6.3}$$

i.e. the vector bundle $J^1(E)$ is the direct sum of

---

[1] If $\Psi \in \Gamma(E)$, recall that $j^1(\Psi)$-th "first order jet" of Ψ - is the cross-section of the vector bundle $J^1(E) \to M$ resulting from defining $j^1(\Psi)(p)$ as the quotient of Ψ in $J^1(E)(p)$ under the mapping:

$$\Gamma(E) \to J^1(E)(p).$$

the vector bundles on the right hand side of (6.3).

Further, $J_{\nabla}^{1}(E)$ maps isomorphically under the bundle projection $J^1(E) \to E$ onto the bundle E. Thus, a connection in E determines a splitting of $J^1(E)$ as a direct sum of $J_h^{1}(E)$ and a bundle which is isomorphic to E.

Proof. Suppose that (Ψ_a), $1 \leq a, b \leq n$, is an F(M)-module basis for $\Gamma(E)$. Let us suppose that:

$$\nabla_X(\Psi_a) = \gamma_{ab}(X)\Psi_b \tag{6.4}$$

for $X \in V(M)$. The $\gamma_{ab}(X)$ - essentially the "Christoffel symbols" - are functions on M.

Then, if:

$$\Psi = f_a\Psi_a, \quad \text{with} \quad f_a \in F(M) \tag{6.5}$$

$$\nabla_X(\Psi) = X(f_a)\Psi_a + f_a\gamma_{ab}(X)\Psi_b.$$

Then, $\nabla_X(\Psi)(p) = 0$ for all $X \in V(M)$ if and only if:

$$X(f_a)(p) + f_a\gamma_{ab}(X)(p) = 0$$

$$\text{for all} \quad X \in V(M). \tag{6.6}$$

In particular, if (x_μ) is a coordinate system for M, if $X = \frac{\partial}{\partial x_\mu} = \partial_\mu$, then:

(6.4) can be written as:

$$\partial_\mu(f_a)(p) + f_a\gamma_{ab\mu}(p), \tag{6.5}$$

where:

$$\gamma_{ab\mu} = \gamma_{ab}(\partial_\mu) \tag{6.6}$$

We see from (6.5) that:

$$J_h{}^1(E)(p) \cap J_\nabla{}^1(E)(p) = 0. \tag{6.7}$$

For if $j^1(\Psi)(p)$ belongs to the left hand side of (6.7), we see that: $f_a(p) = 0$, hence (6.5) implies that $\partial_\mu(f_a)(p) = 0$, i.e. the cross-section Ψ given by (6.3) vanishes to the first order at p, hence: $j^1(\Psi)(p) = 0$.

Second, it should be clear from (6.5) that a Ψ of the form (6.3) can be chosen so that (6.5) is satisfied, while $\Psi(p)$ is an arbitrary vector on E(p). This, together with (6.7), then proves that:

$$J^1(E)(p) = J_h{}^1(E)(p) \oplus J_\nabla{}^1(E)(p). \qquad (6.8)$$

The rest of the statements on Theorem 6.1 follow readily from (6.8).

Now, we can explain the "the principle of minimal interaction". (6.8) implies that a connection ∇ determines an isomorphism between $J^1(E)$ and the direct sum bundle $J_h{}^1(E) \oplus E$. If two connections ∇, ∇' are given, one derives two such isomorphisms, hence an isomorphism $\alpha'_{\nabla,\nabla'}$ of $J^1(E)$ with itself. This isomorphism will be constructable in terms of the "difference" tensor field

$$\tau = \nabla - \nabla'.$$

Before making this more explicit, perhaps it is relevant to discuss in general how such isomorphisms are constructed.

7. ISOMORPHISMS OF JET VECTOR BUNDLES AND THE PRINCIPLE OF MINIMAL INTERACTION

As we have just seen, the "principle of minimal interaction" leads to a linear bundle isomorphism

$$\alpha: \quad J^1(E) \to J^1(E)$$

which is the identity map on the subspace $J_h{}^1(E)(p)$ of $J^1(E)$. If L: $J^1(E) \to R$ is a Lagrangian, one can construct a "new" Lagrangian $\alpha^*(L)$ by the rule:

$$\alpha^*(L)(j) = L(\alpha(j)) \tag{7.1}$$

for $j \in J^1(E)$.

We will now make this more explicit in terms of a local product structure for the bundle, and a coordinization of the base manifold M. Our aim is, of course, to calculate the Euler-Lagrange operators associated with L and L_α.

To keep contact with the case of interest in quantum field theory, suppose M is R^4, with coordinates $x = (x_\mu)$, $0 \le \mu, \nu \le 3$. Let (Ψ_a), $1 \le a, b \le n$, be an F(M)-module basis of $\Gamma(E)$. Set: $\partial_\mu = \frac{\partial}{\partial x_\mu}$.

Now, we recall that $J^1(E)$ is the quotient of $\Gamma(E) \times M$ by a certain equivalence relation. Thus, any real-valued function on $\Gamma(E) \times M$ that is constant on these equivalence classes will "pass to the quotient" to define a function on $J^1(E)$. As examples of this construction, define functions $(\phi_a, \phi_{a\mu})$ as follows:

If $\Psi = f_a \Psi_a$, with $f_a \in F(M)$, then:

$$\begin{aligned} &\text{a)} \quad \phi_a(\Psi, p) = f_a(p) \\ &\text{b)} \quad \phi_{a\mu}(\Psi, p) = \partial_\mu f_a(p) \end{aligned} \qquad (7.2)$$

Thus, somewhat symbolically, we have:

$$\begin{aligned} &\text{a)} \quad \Psi = \phi_a \Psi_a \\ &\text{b)} \quad \phi_{a\mu} = \partial_\mu \phi_a \end{aligned} \qquad (7.3)$$

Now, these real-valued functions, i.e. ϕ_a, $\phi_{a\mu}$, on $\Gamma(E) \times M$ "pass to the quotient" to define functions on $J^1(E)$. We will not use any special notation for this, but hope that the reader can keep things straight.

Suppose now that $L\colon\ J^1(E) \to R$ is a Lagrangian, and that $D_L\colon\ \Gamma(E) \to \Gamma(E)$ is the corresponding Euler-Lagrange differential operator.

If $\Psi \in \Gamma(E)$, and:

$$\phi_a(x) = \phi_a(j^1(\Psi)(x)),$$

then of course:

$$\begin{aligned}&\phi_a(D(\Psi)(x))\\ &= \frac{\partial}{\partial x_\mu} L_{a\mu}(\phi(x), \partial\phi(x)) - L_a(\phi(x), \partial\phi(x)). \quad (7.4)\end{aligned}$$

Our goal now is to compute the operator $D_{\alpha^*(L)}$, where $\alpha^*(L)$ is given by (7.1). Now, $J_h{}^1(E)$ is determined by the following conditions:

$$\begin{aligned}&J_h{}^1(E) = \text{the set of } j \in J^1(E)\\ &\text{such that} \quad \phi_a(j) = 0.\end{aligned}$$

Suppose that:

$$\begin{aligned}\alpha^*(\phi_a) &= \alpha_{ab}\phi_b + \alpha_{ab\mu}\phi_{b\mu}\\ \alpha^*(\phi_{a\mu}) &= \alpha_{a\mu b}\phi_b + \alpha_{a\mu b\nu}\phi_{b\nu}\end{aligned} \quad (7.5)$$

α^* denotes the "dual" map: $F(J^1(E)) \to F(J^1(E))$ on functions induced by the map α. Now, the condition that α maps fibers *linearly* is equivalent to the statement that the coefficients $(\alpha_{ab}, \alpha_{ab\mu}, \alpha_{a\mu b}, \alpha_{a\mu b\nu})$ in (7.5) are *functions of* x *alone*. Let us look for the additional conditions that α leave fixed the elements of $J_h^1(E)$. This condition obviously is that (7.5) take the following special form:

$$\alpha^*(\phi_a) = \alpha_{ab}\phi_b$$
$$\alpha^*(\phi_{a\mu}) = \alpha_{a\mu b}\phi_b + \phi_{a\mu}. \qquad (7.6)$$

Now, as for any function on the manifold $J^1(E)$, we have:

$$dL = L_a d\phi_a + L_{a\mu} d\phi_{a\mu} \qquad (7.7)$$

With $L_\alpha = \alpha^*(L)$, we have:

$$d\alpha^*(L) = \alpha^*(L_a)d\alpha^*(\phi_a)$$
$$+ \alpha^*(L_{a\mu})d\alpha^*(\phi_{a\mu})$$

= , using (7.6).

$$\alpha^*(L_{a\mu})d(\alpha_{ab}\phi_b) + \alpha^*(L_{a\mu})d(\alpha_{a\mu b}\phi_b + \phi_{a\mu})$$
$$= \alpha^*(L_a)\alpha_{ab}d\phi_b + \alpha^*(L_{a\mu})(\alpha_{a\mu b}d\phi_b + d\phi_{a\mu})+\ldots$$

(The terms ... indicate terms in dx_μ; it will not be necessary to be explicit about them.) We can read off from this relation the following rules:

$$(\alpha^*(L))_{a\mu} = \alpha^*(L_{a\mu})$$
$$(\alpha^*(L))_b = \alpha^*(L_a)\alpha_{ab} + \alpha_{a\mu b}\alpha^*(L_{a\mu}) \tag{7.8}$$

Relations (7.8) are what we need. (The use of "differential forms" in (7.7) was not really necessary to derive (7.8), but is presented only as an illustration of their utility.) They tell us how the coefficients in the Euler-Lagrange operator change when L is replaced by $\alpha^*(L)$. In fact, we see from (7.8) that - if $\Psi \in \Gamma(E)$ - then the "Euler-Lagrange differential operator" associated with the "new" Lagrangian (i.e. after a "minimal interaction" sort of change) is:

$$\frac{\partial}{\partial x_\mu}(L_{a\mu}(\alpha(j^1(\Psi)(x))) - \alpha_{ab}(x)L_a(\alpha(j^1(\Psi)(x))$$

$$- \alpha_{b\mu a} L_b(\alpha(j^1(\Psi)(x)). \tag{7.9}$$

For example, suppose that:

$$\begin{aligned}
&\text{a)} \quad L = \beta_{a\mu}(\phi)\phi_{a\mu}. \\
&\text{b)} \quad \alpha_{ab} = \delta_{ab} \\
&\text{c)} \quad \alpha_{a\mu b} = - A_\mu \delta_{ab} \\
&\text{d)} \quad \beta_{a\mu,\, b} = \frac{\partial \beta_{a\mu}}{\partial \phi_b}
\end{aligned} \tag{7.10}$$

Then, the Euler-Lagrange operator associated with L is:

$$\frac{\partial}{\partial x_\mu} (\beta_{a\mu}(\phi(x)) - \beta_{b\mu,\, a}(\phi(x))\partial_\mu \phi_b(x) \tag{7.11}$$

From (7.9), we see that the Euler-Lagrange differential operator associated with $\alpha^*(L)$ is:

$$\frac{\partial}{\partial x_\mu} (\beta_{a\mu}(\phi(x)) - \beta_{b\mu,\, a}(\phi(x))(\partial_\mu - A_\mu)\phi_b(x) \tag{7.12}$$

One recognizes that (7.12) is obtained from (7.11) by what the physicists call a "minimal electromagnetic interaction", i.e. $\partial_\mu \to (\partial_\mu - A_\mu)$, where $A_\mu(x)$ are the "vector potentials" of the electromagnetic field. Thus, we may say that, in the general case, the functions $(\alpha_{ab}(x), \alpha_{a\mu b}(x))$ are the "potentials" of a more complicated field, which has some sort of similarity - by generalization to the electromagnetic field. These more general "fields" are usually called - as some sort of generic name - "Yang-Mills fields".

We also hope that these calculations illustrate a general point to the reader. As we have emphasized, the use of "Lagrangian" in quantum field theory is mainly for the purpose of constructing the corresponding Euler-Lagrange differential operators; the more profound aspects of the calculus of variations do not seem to play a key role. Now, a "Lagrangians" is just a function on a bundle sitting over the base manifold. Thus, any sort of geometric transformation on this bundle will induce a new Lagrangian, thus changing the corresponding differential operator. From a purely mathematical point of view, one could just as readily describe things in terms of transformations on the differential operators themselves.

CHAPTER IV

THE EULER-LAGRANGE OPERATOR IN TERMS OF DIFFERENTIAL FORMS AND JET-BUNDLES

In Chapters 2 and 3, we have discussed the calculus of variations formalism needed to deal with quantum field theory in a form that is intermediate between the "classical" and the "modern", which uses "jet-bundle" ideas. In "Differential geometry and the calculus of variations" and in "Lie algebras and quantum mechanics" we have presented material-following E. Cartan-interrelating the theory of differential forms on manifolds and the calculus of variations. In this chapter, we will proceed further in these directions. It is not really necessary that the reader who is only interested in the main ideas read this chapter seriously, since the main virtue of this approach is to give a very elegant formulation of some of the underlying differential-geometric ideas.

1. DIFFERENTIAL-GEOMETRIC NOTATIONS

Let π: $E \to M$ be a fiber space. $\Gamma(E)$ denotes the space of its cross-sections; $J^r(E)$ denotes the "bundle of r-jets of cross-sections". (Recall that $J^r(E)$ is the quotient of $\Gamma(E) \times M$ by an equivalence relation.)

We will assume - for this chapter - that the reader is familiar with the vector-field - differential-form formalism of differential geometry, as developed in "Differential geometry and the calculus of variations."

It will be convenient for us to work with bases of differential forms for E and M. (What are in differential geometry called "moving frames".) For this purpose, choose the following range of indices and summation conventions:

$1 \leq i, j \leq m = \dim M$

$m+1 \leq a, b \leq n = \dim E.$

Suppose that (ω_i) and (ω_a) are 1-differential forms on M and E, respectively, such that:

$$\begin{aligned} &\text{a)} \quad (\omega_i) \text{ is a basis for 1-forms on } M \\ &\text{b)} \quad (\pi_i^*(\omega_i), \omega_a) \text{ form a basis for 1-forms on } E. \end{aligned} \tag{1.1}$$

(Of course, the reader must keep in mind that "π^*" here denotes the pull-back mapping on functions and differential forms induced by the map π - and not, as in quantum mechanics - complex conjugate or the adjoint operation.) Condition (1.1b) can be replaced by saying that the (ω_a) - when restricted - form a basis for the fibers of the fiber space $\pi\colon E \to M$.

If $\Psi\colon M \to E$ is a cross-section, one can then pull-back the forms ω_a under Ψ, and express them in terms of the ω_i. Say, that:

$$\psi^*(\omega_a) = f_{ai}\omega_i, \tag{1.2}$$

where the (f_{ai}) are functions on M, which depend on Ψ, of course. We can also differentiate the functions (f_{ai}), and express the derivatives in the following form:

$$df_{ai} = f_{aij}\omega_j$$
$$df_{aij} = f_{aijk}\omega_k \qquad (1.3)$$

and so forth.

The functions $(f_{aij}, f_{aijk},\ldots)$ are essentially defined as the functions appearing on the right hand side of (1.3). (Since the ω_i are a "basis" for 1-forms on M, any 1-form on M can be expressed as a linear combination of them, with functions as coefficients. These coefficient functions are what we have defined as the f's.)

Now, these f's are implicitly functions both of the chosen $\Psi \in \Gamma(E)$ and the points of M. We can then regard them as real-valued functions on $\Gamma(E) \times M$. To avoid possible confusion, let us re-label them, using "y" in place of "f" to denote the function - as reinterpreted - on $\Gamma(E) \times M$. Thus, $(y_{ai}, y_{aij},\ldots)$ are real-valued functions on

$\Gamma(E) \times M$ such that:

$$y_{ai}(\Psi, p) = f_{ai}(p)$$
$$y_{aij}(\Psi, p) = f_{aij}(p) \tag{1.4}$$

and so forth.

Notice that some of these functions are constant on the equivalence classes in $\Gamma(E) \times M$ that are used to define $J^r(E)$ as the quotient. For example, if Ψ, Ψ' are elements of $\Gamma(E)$ that agree to the first order at a point $p \in M$, and if:

$$\Psi^*(\omega_a) = f_{ai}\omega_i$$
$$\Psi'^*(\omega_a) = f'_{ai}\omega_i,$$

then: $f_{ai}(p) = f'_{ai}(p)$. (This becomes evident when the ω_i are expressed in terms of a coordinate system for M.) Thus, we have:

$$y_{ai}(\Psi, p) = y_{ai}(\Psi', p).$$

This shows that the (y_{ai}) "pass to the quotient" to define functions on $J^r(E)$, for $r \geq 1$. Since it should not lead to any real confusion, we will make no notational distinction between the symbol "y_{ai}" as functions on $\Gamma(E) \times M$ or as functions on $J^r(E)$.

Similarly, the y_{aij} "pass to the quotient" to define functions on $J^r(E)$, for $r \geq 2$. (But *not* on $J^1(E)$, of course.) y_{aijk} arise from functions on $J^r(E)$, $r \geq 3$, and so forth. For the purposes of the calculus of variations, it is usually sufficient to work with $J^1(E)$ and with the y_{ai} alone, hence we will mainly restrict our attention to this simplest case.

Recall that $J^1(E)$ admits projection maps onto E and M. Using these maps, one can pull back the forms ω_a and ω_i. Again - for the sake of notational simplicity - we will make no notational distinction between these forms on the spaces on which they were originally given and their pull-backs to $J^1(E)$.

With these notations, one readily verifies that the 1-forms on $J^1(E)$:

$$(\omega_i, \omega_a, dy_{ai}) \tag{1.5}$$

form a basis for 1-forms on $J^1(E)$. Many features

of the calculus of variations take a very elegant form when expressed in terms of this basis.

To see the relation of the more concrete way of treating $J^1(E)$ used in previous chapters, let us consider the special case:

$$E = R^m \times R^{n-m},$$

$$\text{with } R^m = M.$$

Let $x = (x_i)$ and $(\phi_a) = \phi$ denote Cartesian coordinates (with the associated vector notations) on these spaces. Thus, an element of $\Gamma(E)$ can be associated with a vector-valued function $x \to \phi(x) = (\phi_a(x))$, with

$$\Psi(x) = (\Psi, \phi(x)). \tag{1.6}$$

Then ω_i, ω_a can be defined as follows:

$$\begin{aligned} \omega_i &= dx_i \\ \omega_a &= d\phi_a \end{aligned} \tag{1.7}$$

Thus, with $\Psi\colon\ M \to E$ defined by (1.6),

$$\Psi^*(\omega_a) = \frac{\partial \phi_a}{\partial x_i} \omega_i, \qquad (1.8)$$

i.e. $f_{ai} = \dfrac{\partial \phi_a}{\partial x_i}$.

Thus,

$$y_{ai}(\Psi, x) = \frac{\partial \phi_a}{\partial x_i}(x) \qquad (1.9)$$

Suppose that L: $J^1(E) \to R$ is a "Lagrangian". In general, define functions L_i, L_a, L_{ai} on $J^1(E)$, using the basis (1.5) for 1-forms on $J^1(E)$, as follows:

$$dL = L_i\omega_i + L_a\omega_a + L_{ai}dy_{ai} \qquad (1.10)$$

Then, with the special choices (1.7), and with (1.9), we see that - when L is expressed as a function $L(x, \phi, y)$ of the indicated variables - the coefficients in (1.10) can be expressed as follows:

$$L_i = \frac{\partial L}{\partial x_i}$$

$$L_a = \frac{\partial L}{\partial \phi_a}$$

$$L_{ai} = \frac{\partial L}{\partial y_{ai}} \left(= \frac{\partial L}{\partial\left(\frac{\partial \phi_a}{\partial x_i}\right)} \right)$$

Thus, the Euler-Lagrange differential operator takes the form:

$$\Psi \to \left[\frac{\partial}{\partial x_i}\left(L_{ai}\left(x, \phi(x), \frac{\partial \phi}{\partial x}\right)\right) - L_a\left(x, \phi(x), \frac{\partial \phi}{\partial x}\right)\right] \tag{1.11}$$

We will see later on in this chapter how (1.11) is described in case the ω's are of more general form than (1.7), e.g. if $d\omega_i \neq 0$ or $d\omega_a \neq 0$. This more general description is of great utility in certain problems where it is not "natural" to choose moving frames of type (1.7). (For example, in "Differential geometry and the calculus of variations" we have described by the Euler-equations associated with a rotating top take a simple form if the "moving frames" are chosen as of more general type.)

2. PROLONGATIONS OF BUNDLE ISOMORPHISMS AND VECTOR FIELDS

Continue with the notations for (E, M, π) and

$J^1(E)$ introduced in Section 1. Suppose that α: $E \to E$ is a diffeomorphism of E that is also a bundle isomorphism. Recall that this means that there is a "base" diffeomorphism α_M: $M \to M$ which is "covered" by α, i.e.

$$\pi\alpha = \alpha_M\pi$$

This means, explicitly, that:

$$\alpha(E(p)) = E(\alpha_M p)$$

for $p \in M$.

As we know, α acts on cross-sections:

$$(\alpha\Psi)(p) = \alpha(\Psi(\alpha^{-1}p)) \tag{2.1}$$

for $p \in M$.

Thus, (2.1) defines α as a transformation: $\Gamma(E) \to \Gamma(E)$. We can then extend α to a transformation: $\Gamma(E) \times M \to \Gamma(E) \times M$ using the following

$$\alpha(\Psi, p) = (\alpha\Psi, \alpha_M p)$$

for $\Psi \in \Gamma(E)$, $p \in M$.

This action of α on $\Gamma(E) \times M$ clearly preserves the equivalence classes of the equivalence relation whose quotient is $J^r(E)$. Therefore, the action of α again "passes to the quotient" to define a transformation α^r: $J^r(E) \to J^r(E)$. α^r is called the *r-th order prolongation* of α. Two facts must be verified (but are routine):

a) α^r is a diffeomorphism with respect to the manifold structure defined for $J^r(E)$.

b) If $\alpha = \alpha_1\alpha_2$, where α_1, α_2 are bundle isomorphisms, then

$$\alpha^r = \alpha_1{}^r\alpha_2{}^r. \tag{2.2}$$

Thus, (2.2) tells us that a given group, G, of bundle diffeomorphisms "prolongs" to a group G^r of bundle diffeomorphisms of $J^r(E)$.

We must now examine this "prolongation" process - which is so simple geometrically on the "group" level just explained - at the "infinitesimal" or "Lie algebra" level. To do this, suppose that $t \to \alpha_t$ is a one parameter group of bundle isomorphisms. Then, the corresponding "base" diffeomorphism $\alpha_{M,\ t}$: $M \to M$ also form a one-parameter groups, have - as for any one-parameter transformation group - "infinitesimal generators" X and X_M, which are vector fields on E and M.

Recall their definitions:

$$X(f) = \frac{\partial}{\partial t} \alpha_{-t}^{*}(f)/_{t=0}$$

for $f \in F(E)$. (2.3)

$$X_M(f) = \frac{\partial}{\partial t} \alpha_{M,\ -t}^{*}(f)/_{t=0}$$

for $f \in F(M)$.

Our goal is to compute the analogously defined "infinitesimal generator" of the "prolonged" one-parameter group $t \to \alpha_t{}^r$, which will be a vector field, denoted by X^r, on E. For simplicity (and, because it is the only case really needed in most calculus of variations problems), we will deal only with the case: $r = 1$:

Suppose that (ω_i, ω_a) is the basis for 1-forms on E utilized in Section 1. (By "computing" X^1 we mean that we want to express it in terms of this basis and the functions $y_{ai} \in F(J^1(E))$ defined in Section 1.) Then, the assumptions made so far imply that Lie derivative by X effects this basis of 1-forms in the following way:

$$X(\omega_i) = A_{ij}\omega_j$$
$$X(\omega_a) = A_{ai}\omega_i + A_{ab}\omega_b \tag{2.4}$$

If one regards the ω_i and ω_a as forms on $J^1(E)$ - via the pull-back π^{1*} - then the relations (2.3) hold, with X^1 replacing X. Thus, to determine X^1 - in terms of its Lie derivative action on forms on $J^1(E)$ - it only remains to calculate the Lie derivative of the y_{ai} by X^1.

To do this, it is most convenient to proceed somewhat indirectly, and think about the geometry. Introduce the following set of 1-forms on $J^1(E)$:

$$\theta_a = \omega_a - y_{ai}\omega_i. \tag{2.5}$$

Now, examining the procedure used to *define* the functions y_{ai}, we see that the following property holds:

If Ψ: $M \to E$ is a cross-section of the

fiber space, and if $j^1(\Psi)\colon M \to J^1(E)$ is its "1-jet", then:

$$j^1(\Psi)^*(\theta_a) = 0. \tag{2.6}$$

In terms of the jargon of differential geometry, the 1-jets are "integral submanifolds" of the Pfaffian system determined by the θ_a. Now, if $\alpha^1\colon J^1(E) \to J^1(E)$ is a "prolongation" of a bundle isomorphism of E, it maps these 1-jet cross-sections into 1-jet cross-sections. This indicates that α^{1*} must preserve the system (θ_a) of 1-forms, i.e. there are relations of the form:

$$\alpha^{1*}(\theta_a) = f_{ab}\theta_b, \tag{2.7}$$

with (f_{ab}) functions on $J^1(E)$.

Now, when one-parameter groups $t \to \alpha_t{}^1$ of such isomorphisms are considered, and the partial derivative of (2.7) with respect to t is taken, relations of the following form are obtained:

$$X^1(\theta_a) = f_{ab}\theta_b, \tag{2.8}$$

where (f_{ab}) are functions on $J^1(E)$ (not the same as in (2.7), of course). Now one can readily verify - a task we leave as an exercise - directly from the definition of X^1 given above that relations of the form (2.8) hold. We will now show how the coefficients f_{ab} in (2.8) are determined in terms of (2.6) and the coefficients on the right hand side of (2.4).

Suppose then that, also,

$$X^1(\omega_i) = A_{ij}\omega_j$$
$$X^1(\omega_a) = A_{ai}\omega_i + A_{ab}\omega_b. \tag{2.9}$$

Using (2.5), we have:

$$X^1(\theta_a) = A_{ai}\omega_i + A_{ab}\omega_b$$
$$- X^1(y_{ai})\omega_i - y_{ai}A_{ij}\omega_j \tag{2.10}$$

= , also, using (2.8) and (2.5),

$$f_{ab}(\omega_b - y_{bi}\omega_i).$$

Comparing coefficients, we have:

$$f_{ab} = A_{ab} \tag{2.11}$$

$$A_{ai} - X^1(y_{ai}) - y_{aj}A_{ji} = - f_{ab}y_{bi},$$

or, using (2.11),

$$X^1(y_{ai}) = A_{ai} - y_{ji} + A_{ab}y_{bi}. \tag{2.12}$$

Combining (2.11) and (2.8) gives:

$$X^1(\theta_a) = A_{ab}\theta_a. \tag{2.13}$$

Formulas (2.12) and (2.13) are the basic formulas we will need to work on the calculus of variations. From then, one can readily derive the following formula:

$$[X, Y]^1 = [X^1, Y^1] \tag{2.14}$$

if $X, Y \in V(E, M)$ are vector fields on E that are infinitesimal generators of one-parameter groups of fiber space automorphisms.

Of course, (2.14) is implicit in the group-theoretic-geometric explanation of the prolongation operation, since, if G is a group of fiber space automorphisms acting on E, and if G^1 is its prolongation, acting on $J^1(E)$, the mapping $g \to g^1$ of $G \to G^1$ is a group homomorphism. $X \to X^1$ is then the Lie algebra homorphism (which is what (2.14) says, of course) obtained by passing to the Lie algebra.

3. THE CALCULUS OF VARIATIONS IN TERMS OF JET-BUNDLES

E, M, π, $J^1(E)$, $(\omega_i, \omega_a, y_{ai}, \theta_a)$ will be as defined in Sections 1-2. Let L be a real-valued function on $J^1(E)$, i.e. a "Lagrangian". Let dp be a volume-element differential form (of degree m = dim M) on M. For simplicity, we will suppose that dp is related to the (ω_i) by the following formula:

$$dp = \omega_1 \wedge \cdots \wedge \omega_m \tag{3.1}$$

($\wedge$ denotes the "exterior product" operation for differential forms).

Define a real-valued function $\underset{\sim}{L}$: $\Gamma(E) \to R$

as follows:

$$\underset{\sim}{L}(\Psi) = \int_M j^1(\Psi)^*(L)dp. \tag{3.2}$$

Suppose now that $t \to \alpha_t$ is a one-parameter group of fiber space automorphisms acting on E, such that:

$$\alpha_t(E(p)) = E(p) \tag{3.3}$$

for all $p \in M$,

i.e. each α_t is a fiber space isomorphism of E that acts "trivially" on the base M, i.e. preserves the fibers. Let X denote the vector field on E which is its infinitesimal generator. One readily sees that the following condition characterizes such X's:

$$X(\pi^*(f)) = 0 \quad \text{for all} \quad f \in F(M) \tag{3.4}$$

The following notation will be helpful: Let $V_o(E, M)$ denote the space of vector fields X on E satisfying (3.4). Define V(E, M) as follows:

A vector field $X \in V(E)$ is in $V(E, M)$ if there is a vector field $X_M \in V(M)$ such that:

$$X(\pi^*(f)) = \pi^*(X_M(f)) \tag{3.5}$$

for all $f \in F(M)$

In terms of differential-geometric jargon, $V(E, M)$ consists of the vector fields on E that are "projectable" under π, with $X_M \in V(M)$ its "projection". One readily verifies that the map $X \to X_M$ of $V(E, M) \to V(M)$ is a Lie algebra homomorphism. $V_o(E, M)$ then consists of the $X \in V(E, M)$ such that:

$$X_M = 0.$$

One readily verifies that:

$$[V(E, M), V_o(E, M)] \subset V_o(E, M), \tag{3.6}$$

i.e. $V_o(E, M)$ is an *ideal* in the Lie algebra $V(E, M)$.

Suppose now again that α_t satisfies (3.3).

Set:

$$\Psi^{\lambda} = \alpha_{\lambda}(\Psi) \tag{3.7}$$

(We have changed the label from "t" to "λ" to conform with the notation used in Chapter 2.) Our goal now is to compute the "First variation":

$$\frac{d}{d\lambda} L(\Psi^{\lambda})/_{\lambda=0} \tag{3.8}$$

and from this to compute the "Euler-Lagrange differential operator" generated by X.

Suppose that:

$$dL = L_i \omega_i + L_a \omega_a + L_{ai} dy_{ai}. \tag{3.9}$$

Now, from the geometric meaning of the "prolongation" operation, we have:

$$L(\Psi^{\lambda}) = \int_M (\alpha_{\lambda}{}^1(j^1(\Psi)))^*(L)dp$$

$$= \int_M j^1(\Psi)(\alpha_{\lambda}{}^{1*}(L))dp,$$

hence:

$$\frac{d}{d\lambda} L(\Psi^{\lambda})/_{\lambda=0} = \int_M j^1(\Psi)^*(X^1(L))dp. \qquad (3.10)$$

Thus, using (3.9) and (2.12) we have immediately available a formula for (3.10) in terms of the function (A_{ai}, A_{ab}) in (2.9). Notice that the conditions $X \varepsilon V_o(E, M)$ implies that:

$$\omega_i(X) = X(\omega_i) = 0,$$

hence

$$A_{ij} = 0. \qquad (3.11)$$

Set:

$$A_a = \omega_a(X). \qquad (3.12)$$

What is desired is a formula for (3.10) in terms of the functions A_a and *not its derivatives*. (The A_{ai}, A_{ab} will involve its derivatives.) Thus, we must "integrate by parts", i.e. apply Stokes' theorem, to the right hand side of (3.10), assuming that the "boundary terms" vanish. To do this, let us make (3.10) explicit, in terms (3.9): Now, from (2.12) and (3.11) imply that:

$$X^1(y_{ai}) = A_{ai} - A_{ab}y_{bi}.$$

From (3.9) and (3.12)

$$X^1(L) = L_a A_a + L_{ai}(A_{ai} - A_{ab}y_{bi}). \qquad (3.13)$$

Thus,

$$\frac{d}{d\lambda} \underset{\sim}{L}(\Psi^\lambda)/_{\lambda=0}$$

$$= \int_M j^1(\Psi)^*(L_a A_a + L_{ai}(A_{ai} - A_{ab}y_{bi}))\omega_1 \wedge \cdots \wedge \omega_m. \qquad (3.14)$$

Introduce (m-1)-forms θ_i on $J^1(E)$ as follows:

$$\begin{aligned} \theta_1 &= \omega_2 \wedge \cdots \wedge \omega_m \\ \theta_2 &= -\omega_1 \wedge \omega_3 \wedge \cdots \wedge \omega_m \\ &\vdots \\ \theta_m &= \pm\omega_1 \wedge \cdots \wedge \omega_{m-1}. \end{aligned} \qquad (3.15)$$

(The general formula is:

$$\theta_i = (-1)^{i-1}\omega_1 \wedge \cdots \wedge \hat{\omega}_i \wedge \cdots \omega_m,$$

where the notation ^ above a form means that it is to be omitted.)

Then,

$$\begin{aligned} & L_{ai}(A_{ai} - A_{ab}y_{bi})\omega_1 \wedge \cdots \wedge \omega_m \\ &= L_{ai}(A_{aj} - A_{ab}y_{aj})\omega_j \wedge \theta_i \qquad (3.16) \\ &= L_{ai}A_{aj}\omega_j \wedge \theta_i + L_{ai}A_{ab}(\theta_b + \omega_b) \wedge \theta_i . \end{aligned}$$

Now, $j^1(\Psi)^*(\theta_b) = 0$, hence the term in (3.16) involving θ_b can be ignored in computing (3.14).

Also,

$$X(\omega_a) = A_{ab}\omega_b + A_{ai}\omega_i,$$

hence, (3.14) can be rewritten as:

$$\begin{aligned} & \frac{d}{d\lambda} \underset{\sim}{L}(\Psi^\lambda)/_{\lambda=0} \\ &= \int_M j^1(\Psi)^*(L_a A_a dp + L_{ai}X(\omega_a)\theta_j). \qquad (3.17) \end{aligned}$$

Suppose now that:

$$d\omega_a = h_{bca}\omega_b \wedge \omega_c + h_{bia}\omega_b \wedge \omega_i + h_{ija}\omega_i \wedge \omega_j, \tag{3.18}$$

where the h's are functions on E. Now,

$$X(\omega_a) = d(X \lrcorner\, \omega_a) + X \lrcorner\, d\omega_a$$

$$= dA_a + h_{bca}A_b\omega_c + h_{bi}A_b\omega_i.$$

Thus, we can rewrite (3.17) as:

$$\frac{d}{d\lambda} \underset{\sim}{L}(\Psi^\lambda)/_{\lambda=0}$$

$$= \int_M j^1(\Psi)^*(A_a[L_a dp - d(L_{ai}\theta_i) + L_{bi}(h_{acb}\omega_c + h_{aj}\omega_j) \wedge \theta_i] + d(L_{ai}A_a\theta_i)). \tag{3.19}$$

The last term on the right hand side of (3.10) will vanish *if* Stokes' theorem is applied, and - as an Ansatz - $\Psi^*(A_a)$ vanish at the "boundary" of M, resulting in the definitive form of the "first variation formula":

$$\frac{d}{d\lambda} \underset{\sim}{L}(\Psi^{\lambda})/_{\lambda=0}$$

$$= \int_M j^1(\Psi)^*(A_a[L_a dp - d(L_{ai}\theta_i) \tag{3.20}$$

$$- L_{bi}\theta_i \wedge (h_{acb}\omega_c + h_{aj}\omega_j)].$$

One can read off from this the "Euler-Lagrange differential operator":

$$\Psi \to j^1(\Psi)^*[L_a dp - d(L_{ai}\theta_i)$$

$$- L_{bi}\theta_i \wedge (h_{acb}\omega_c + h_{aj}\omega_j]. \tag{3.21}$$

For example, suppose that (x_i, ϕ_a) are functions on E, with:

$$\omega_j = dx_i; \ \omega_a = d\phi_a$$

$$\Psi^*(\phi_a) = \phi_a(x). \tag{3.22}$$

Then, $h_{acb} = 0 = h_{aj}$, and (3.21) takes the more familiar form:

$$\Psi \to j^1(\Psi)^*(L_a dx) - \left[\frac{\partial}{\partial x_i} j^1(\Psi)^*(L_{ai})\right]dx, \tag{3.23}$$

with $dx = dp = dx_1 \wedge \dots \wedge dx_m$.

Thus, formula (3.21) has the virtue of expressing the Euler-Lagrange differential operator in terms of more general "moving frames" than the "flat" ones, of form (3.22). As we shall see later, certain variational problems that are of possible interest as candidates for "quantization" (and, as description of "elementary particles") can be more readily described in terms of "non-flat moving frames".

Now, we turn to describing the connection with Cartan's "differential form" description of the calculus of variations.

4. THE EULER-LAGRANGE OPERATOR IN TERMS OF THE CARTAN VARIATIONAL FORMALISM

Continue with the notations of Section 3. In "Lie algebras and quantum mechanics" we have described how these "first variational formulas" can be elegantly derived in terms of differential forms. We will now show how they adapt to the situation of Section 3.

Introduce m-differential forms Ω_1, Ω_2, Ω on

$J^1(E)$, as follows:

$$\Omega_1 = L\,dp$$

$$\Omega_2 = L_{ai}\theta_a \wedge \theta_i \tag{4.1}$$

$$\Omega = \Omega_1 + \Omega_2.$$

Then, since $j^1(\Psi)^*(\theta_a) = 0$, we have:

$$\underset{\sim}{L}(\Psi) = \int_M j^1(\Psi)^*(\Omega). \tag{4.2}$$

Thus,

$$\frac{d}{d\lambda}\underset{\sim}{L}(\Psi^\lambda)/_{\lambda=0} = \int_M j^1(\Psi)^*(X^1(\Omega)) \tag{4.3}$$

$$= \int_M j^1(\Psi)^*(X^1 \lrcorner\, d\Omega + d(X^1 \lrcorner\, \Omega)).$$

Again, assume that the second term on the right hand side of (4.3) vanishes after Stokes' theorem is applied, resulting in the following formula:

$$\frac{d}{d\lambda}\underset{\sim}{L}(\Psi^\lambda)/_{\lambda=0} = \int_M j^1(\Psi)^*(X^1 \lrcorner\, d\Omega). \tag{4.4}$$

Now, a-priori, the m-form $X^1 \lrcorner\, d\Omega$ depends on the *derivatives* of the components of the vector field X. The remarkable property of Ω is that the terms involving the derivatives vanish in (4.4). Explicitly, one has the following formula:

$$\begin{aligned} &\psi^{1*}(X^1 \lrcorner\, d\Omega) \\ &= \psi^{1*}(A_a L_a dp - A_a d(L_{ai}\theta_j) \\ &+ L_{ai}(X \lrcorner\, d\omega_a) \wedge \theta_j). \end{aligned} \qquad (4.5)$$

Thus, we have proved a main result:

<u>THEOREM 4.1</u>. $\Psi \in \Gamma(E)$ is an extremal of the variational problem defined by the Lagrangian L if and only if:

$$\psi^{1*}(V_o(E, M) \lrcorner\, d\Omega) = 0. \qquad (4.6)$$

The Euler-Lagrange differential operator can now be defined as follows. $V_o(E, M)$ is an $F(M)$-module. Let Γ denote the space all linear maps γ: $V_o(E, M) \to$ (space of m-forms on M) such that:

$$f\gamma(X) = \gamma(\pi^*(f)) \quad \text{for} \quad f \in F(M). \tag{4.6}$$

Then, define D_L: $\Gamma(E) \to \Gamma$ by the rule:

$$D_L(\Psi) = \gamma, \tag{4.7}$$

with:

$$\gamma(X) = j^1(\Psi)^*(X \lrcorner d\Omega)$$

$$\text{for all} \quad X \in V_o(E, M).$$

(4.5) then shows explicitly that (4.6) is satisfied. If, then, (E', M, π') is a fiber space over M, whose cross-sections $\Gamma(E')$ can be identified with Γ, D_L would map: $\Gamma(E) \to \Gamma(E')$, i.e. the "Euler-Lagrange differential operator" D_L would be identified with a differential operator (in general, "non-linear", of course) in the geometric sense. Since we will not need this interpretation in this work, we will not pursue it here, although it may be useful in later work.

CHAPTER V

THE SYMPLECTIC STRUCTURE ON THE SPACE OF EXTREMALS; APPLICATION TO QUANTUM FIELD THEORY

1. INTRODUCTORY REMARKS

Our approach to quantum field theory will combine ideas of two of the great masters of mathematics and physics in the 20th century, E. Cartan and P. Dirac. What we want to do is to give some idea of the geometric principles underlying the "quantization" idea.

Suppose that M is a space that shares with the class of differentiable manifolds certain simple geometric properties. (We have in mind M being something like an "infinite dimensional differentiable manifold", but will not use this term in the precise technical sense used in the literature of "global analysis".)

We will suppose that M has three "primitive" geometric structures:

a) An algebra (under point-wise multiplication) of real valued functions (thought of as the "C^∞ functions") denoted by $F(M)$.

b) A "tangent" vector space attached to each point $p \in M$, denoted by M_p. Each $v \in M_p$ determines a linear mapping: $F(M) \to R$ such that:

$$v(f_1 f_2) = v(f_1) f_2(p) + f_2(p) v(f_1)$$

for $f_1, f_2 \in F(M)$.

c) A notion of "C^∞ curve" in M; a typical curve denoted by $t \to \sigma(t)$, $-\infty < t < \infty$, such that its "tangent vector" $\sigma'(t) \in M_{\sigma(t)}$ can be defined at each point of σ, with

$$\sigma'(t)(f) = \frac{d}{dt} f(\sigma(t)).$$

Following Dirac, a "Poisson bracket" operation can be defined by giving the following data on M:

a) A subring F of F(M).

b) A Lie algebra operation, denoted by

$$(f_1, f_2) \to \{f_1, f_2\},$$

such that:

$$\{f_1, f_2, f_3\} = \{f_1, f_2\}f_3 + \ldots$$
$$\text{for} \quad f_1, f_2, f_3 \in F. \tag{1.1}$$

"Quantization" means performing the following operations:

a) Selecting a Lie subalgebra of F, denoted typically by S, whose elements are to play the role of the "physical observables".

b) Finding an irreducible linear representation of S by skew-Hermitian operators on a Hilbert space H, with the vectors in H playing the role of the "states" of the system.

Now, precisely how to select the Lie algebra S is unknown, in general. (In fact making precise how S is to be chosen is one of the key problems in understanding the mathematical mysteries inherent - despite forty years of research - in quantum mechanics.) In systems that correspond to the motion of a finite number particles, one usually describes S by the following process:

a) Find a set, (Q_i, P_j), $1 \leq i, j \leq n$, of "canonically conjugate" elements of F, i.e. elements such that

$$\begin{aligned} \{P_i, Q_j\} &= \delta_{ij} \\ 0 = \{P_i, P_j\} &= \{Q_i, Q_j\}. \end{aligned} \tag{1.2}$$

(Thus, the elements 2n+1 - $(P_i, Q_j, 1)$ span a Lie algebra. The "abstract" Lie algebra with those relations as structure relations is called the "Heisenberg" Lie algebra; the corresponding "commutation relations" are called the "Heisenberg commutation relations".)

b) Find an irreducible representation of the P_i, Q_j by skew-Hermitian operators in a Hilbert space H.

c) Try to figure out - for example, using physical intuition - which "classical functions" f(P,Q) of the P's and Q's should be included in the observables S, e.g. the "energy", "angular momentum", etc.

d) Try to extend the representation given by b) (which is essentially uniquely determined once Planck's constant is fixed) to the S's given by c), using - as far as possible - the "Schrödinger" rules of quantization.

We shall not attempt further precision about this process, because basically none is available.

However, suppose one asks: What is the differential-geometric mechanism that *defines* the subalgebra F and the Poisson-bracket operation { , } that starts the ball rolling. In "Lie algebras and quantum mechanics" we have described this in some detail. For our immediate purposes, it suffices to recall a few general, vague remarks.

Suppose given a 2-differential form ω on M.

Now, such an object may be defined as a bilinear function $(v_1, v_2) \to \omega(v_1, v_2)$ on pairs of tangent vectors to points of M, such that:

$$\omega(v_1, v_2) = -\omega(v_2, v_1)$$

$$\text{for} \quad v_1, v_2 \in M_p, \text{ all } p \in M.$$

We will also suppose that ω is "closed", i.e. $d\omega = 0$, where d is the exterior product operation. (Of course, how to define "d" for the general sort of spaces we are considering is another problem, which we will not go into here.)

Now, given such an ω, we can define, first, a subalgebra F(ω) of F(M), and, second, a Poisson-bracket type of Lie algebra operation on F(ω). We have described this in detail in "Lie algebras and quantum mechanics", and it will not be repeated here, at this point at least. (Recall that F(ω) consists of functions that are constant on the characteristic curves of ω.) Instead, we will turn to a description of the 2-form in the case where M is the "space" of all "extremals" of a variational problem.

2. THE SYMPLECTIC[1] STRUCTURE ASSOCIATED WITH A VARIATIONAL PROBLEM

Let us adopt the explicit, classical description of a variational problem. Thus, the fiber space E is $R^4 \times R^n$, with $\pi\colon\ E \to R^4$ the Cartesian product projection. (Since this is the main case of interest in quantum field theory, we will limit ourselves to the case where the base space is four dimensional, for which we adopt the index conventions that are now standard in quantum field theory.) Denote the coordinates of M by

$x = (x_\mu)$,

$$0 \leq \mu, \nu \leq 3, \text{ with } \partial_\mu = \frac{\partial}{\partial x_\mu},$$

coordinates of R^n by $\phi = (\phi_a)$,

$$1 \leq a, b \leq n.$$

---

[1]By a "symplectic structure" on a manifold as mean the geometric structure defined by a given closed, 2-differential form on the manifold. If this closed 2-form has no non-zero characteristic vectors, we will call it a "canonical structure".

Thus, an element of $\Gamma(E)$ can be identified with a vector-valued function $x \to \phi(x) = (\phi_a(x))$. As we have seen, the coordinates of $J^1(E)$ can be chosen as $(x, \phi, \dot{\phi})$, where $\dot{\phi}$ stands for variables labeled $(\phi_{a\mu})$. If $x \to (x)$ is identified with a cross-section of E, its 1-jet can be identified with the map $x \to (x, \phi(x), \partial\phi(x))$, of $M \to J^1(E)$, where $\partial\phi(x)$ stands for:

$$\phi_{a\mu}(x) = \partial_\mu \phi_a(x).$$

Suppose that L: $J^1(E) \to R$ is a Lagrangian. We can then consider L as a function $L(x, \phi, \dot{\phi})$ of the indicated variables.

Introduce the following notations:

$$L_\mu = \frac{\partial L}{\partial x_\mu} \; ; \; L_a = \frac{\partial L}{\partial \phi_a} \; ; \; L_{a\mu} = \frac{\partial L}{\partial \phi_{a\mu}} \; .$$

$$L_{a,b} = \frac{\partial^2 L}{\partial \phi_a \partial \phi_b} \; ; \quad L_{a\mu,b\nu} = \frac{\partial^2 L}{\partial \phi_{a\mu} \partial \phi_{b\nu}} \; ,$$

and so forth.

Let M be the space of all maps: $x \to \phi(x)$: i.e. elements of $\Gamma(E)$, satisfying the Euler-

Lagrange equations:

$$\partial_\mu(L_{a\mu}(x, \phi(x), \partial\phi(x))) = L_a(x, \phi(x), \partial\phi(x)). \tag{2.1}$$

In accordance with the ideas outlined in Section 1, we want to construct the "tangent space" to M at a "point" ϕ. Suppose then that $\lambda \to (\phi_a{}^\lambda(x))$ is a one-parameter family of solutions of (2.1), reducing to ϕ at $\lambda = 0$. Set:

$$\gamma_a(x) = \frac{\partial}{\partial\lambda} \phi_a{}^\lambda(x)/_{\lambda=0}. \tag{2.2}$$

Then, $\gamma = (\gamma_a(x))$ may be considered as a "tangent vector" to M at ϕ, i.e. an element of M_ϕ. We obtain the differential equations defining M_ϕ by differentiating (2.1) with respect to λ. The resulting equations are called the *linear variational equations* associated with the Euler-Lagrange equations (2.1).

Explicitly:

$$0 = \frac{\partial}{\partial\lambda} [\partial_\mu(L_{a\mu}(x, \phi^\lambda(x), \partial\phi^\lambda(x))$$

$$- L_a(x, \phi^\lambda, \partial\phi^\lambda)]/_{\lambda=0}$$

$$= \partial_\mu(L_{a\mu,b}\gamma_b + L_{a\mu,b\nu}\partial_\nu\gamma_b)$$

$$- L_{a,b}\gamma_b - L_{a,b\nu}\partial_\nu\gamma_b. \qquad (2.3)$$

(It should be understood of course that, for example, $L_{a\mu,b}$ stands for: $L_{a\mu,b}(x, \phi(x), \partial\phi(x))$.) With $\phi(x)$ given, (2.3) is then a system of second order, linear differential equations for the $\gamma_a(x)$.

For example, consider a simple special case:

$$L = \frac{1}{2}(g_{\mu\nu}\phi_{a\mu}\phi_{a\nu} + m^2\phi_a\phi_a), \qquad (2.4)$$

where the $(g_{\mu\nu})$ are constants, symmetric in μ and ν. (Thus, if $(g_{\mu\nu})$ is the Lorentz metric tensor, i.e.

$$g_{\mu\nu} = 0 \text{ if } \mu \neq \nu;$$

$$g_{oo} = 1,\ g_{\mu\mu} = -1 \quad \text{for} \quad 1 \leq \mu \leq 3,$$

then L would be Lagrangian for a set of spin zero particles of mass m, e.g. "pions".) In this case:

$$L_{a\mu} = g_{\mu\nu}\phi$$

$$L_a = m^2\phi_a.$$

(2.1) then takes the form:

$$g_{\mu\nu}\partial_\mu\partial_\nu\phi_a = m^2\phi_a. \tag{2.5}$$

(2.5) is then just the "Klein-Gordon equation". Due to its simple linear, constant-coefficient form, the linear variational equations are identical:

$$g_{\mu\nu}\partial_\mu\partial_\nu\gamma_a = m^2\gamma_a. \tag{2.6}$$

Suppose, now, that γ_a, γ'_a are two solutions of (2.6). Form:
Set:

$$\omega_\mu(\gamma, \gamma') = g_{\mu\nu}(\gamma_a\partial_\nu(\gamma'_a) - \gamma'_a\partial_\nu(\gamma_a)) \tag{2.7}$$

Then,

$$\partial_\mu\omega_\mu = g_{\mu\nu}(\partial_\nu\gamma_a\partial_\nu\gamma'_a - \partial_\mu\gamma'_a\partial_\nu\gamma_a - \gamma'_a\partial_\mu\partial_\nu\gamma_a) \tag{2.8}$$

$$= \text{, using (2.6), } 0.$$

Thus, ω_μ is a "conserved current". Explicitly, introduce the following notations:

$$\vec{x} = (x_i),\ 1 \leq i,\ j \leq 3.$$

$$t = x_o;\ x = (t, \vec{x}).$$

$$d\vec{x} = dx_1 \wedge dx_2 \wedge dx_3.$$

Set:

$$\omega(\gamma, \gamma') = \int \omega_o(\gamma(t, \vec{x}), \gamma'(t, \vec{x}))d\vec{x}. \quad (2.9)$$

The right hand side is then independent of t, and defines a skew-symmetric, bilinear form on M_ϕ. This is the desired "symplectic structure" on M_ϕ.

More generally, one can interpret the "conserved current condition" (2.8) as follows: Let $\theta(\gamma, \gamma')$ be the following 3-differential form on $M = R^4$:

$$\theta = \omega_o dx_1 \wedge dx_2 \wedge dx_3 - \omega_1 dx_o \wedge dx_2 \wedge dx_3$$
$$+ \omega_2 dx_o \wedge dx_1 \wedge dx_3 - \omega_3 dx_o \wedge dx_1 \wedge dx_2. \quad (2.10)$$

If N is a 3-dimensional submanifold of $M = R^3$, set:

$$\omega_N(\gamma, \gamma') = \int_N \theta. \tag{2.11}$$

(2.8) is equivalent to the following condition:

$$d\theta = 0. \tag{2.12}$$

Then, for N held fixed, this is a skew-symmetric, bilinear function on M_ϕ. The value is the same for two submanifolds N, N' which together "cobound" a 4-dimensional region of R^4, by (2.12) and Stokes' theorem. This is the "conserved current" condition.

Let us consider another special case:

$$L = h_{a\mu b}\phi_{a\mu}\phi_b + m_{ab}\phi_a\phi_b, \tag{2.13}$$

where the coefficients $(h_{a\mu b})$, (m_{ab}) are constants, with $m_{ab} = m_{ba}$. Then,

$$L_a = h_{a\mu b}\phi_b$$

$$L_a = m_{ab}\phi_b + h_{b\mu a}\phi_{b\mu}.$$

The Euler-Lagrange equations are then:

$$h_{a\mu b}\partial_\mu\phi_b = m_{ab}\phi_b + h_{b\mu a}\partial_\mu\phi_b,$$

or

$$(h_{a\mu b} - h_{b\mu a})\partial_\mu\phi_b = m_{ab}\phi_b. \quad (2.14)$$

Again, since the equations (2.14) are linear, constant coefficient, they are identical with their linear variational equations:

$$(h_{a\mu b} - h_{b\mu a})\partial_\mu\gamma_b = m_{ab}\gamma_b. \quad (2.15)$$

If γ, γ' are two solutions of (2.15), set:

$$\omega_\mu(\gamma, \gamma') = (h_{a\mu b} - h_{b\mu a})\gamma_a\gamma'_b. \quad (2.16)$$

Then,

$$\partial_\mu\omega_\mu = (h_{a\mu b} - h_{b\mu a})(\partial_\mu\gamma_a\gamma'_b + \gamma_a\partial_\mu\gamma'_b)$$

$$= \gamma_a m_{ab}\gamma'_b - \gamma'_b = 0. \quad (2.17)$$

Again, this implies that $\omega_\mu(\gamma, \gamma')$ is a "conserved current", and (2.9) or (2.11) define a 2-differential form on M. With these two examples in hand, rather than treat the general case (2.4) directly, it is more convenient to study the problem from the general point of view of vector bundles.

3. CONSERVED CURRENTS ASSOCIATED WITH LINEAR DIFFERENTIAL OPERATORS

It will be convenient to work with general, vector bundle notions. Suppose (E, M, π) and (E', M, π') are two vector bundles over a manifold M. (The case $M = R^4$, used in the last section, should be kept in mind.) Suppose $D\colon \ \Gamma(E) \to \Gamma(E')$ is a linear differential operator. Our problem:

> How can one define a "conserved current" associated with two solutions $\Psi, \Psi' \in \Gamma(E)$ differential equations:

$$D(\Psi) = 0 = D(\Psi') = 0. \tag{3.1}$$

Looking at the examples worked out in Section 2, we can describe the general pattern that should hold.

Let $T(M)$ be the tangent bundle for M. A cross-section of $T(M)$ is identified in the usual way with the space of "vector fields" on M, i.e. the space V(M) of first order, linear, homogeneous differential operators: $F(M) \to F(M)$. Now, "dp" denotes a fixed, volume element differential form on M, i.e. form of degree $m = \dim M$. Using "dp", we can define a first-order differential operator- denoted by "div" - from $\Gamma(T(M)) = V(M)$ to F(M). This can be done in either of two ways:

a) If $X \in V(M)$, div (X) is the function in F(M) such that

$$X(dp) = (\text{div } X)dp. \tag{3.2}$$

(Recall that "X(dp)" denotes the Lie derivative of dp by X.)

b) If $X \in V(M)$, let θ_X denote the (m-1)-form such that:

$$\theta \wedge \theta_X = \theta(X)dp \tag{3.3}$$

for all 1-forms θ on M.

(θ_X is uniquely determined by (3.3), and is said to correspond to X "by duality".) Then,

$$d\theta_X = (\text{div } X)dp. \tag{3.4}$$

Of course, the function "div X" is called the *divergence of the vector field X*. More material on this operation can be found in "Differential geometry and the calculus of variations". In particular, it is shown there that, if $M = R^4$, with coordinates $x = (x_\mu)$, $0 \le \mu, \nu \le 3$, and if:

$X = A_\mu \partial_\mu$, then:

$\operatorname{div} X = \partial_\mu (A_\mu)$.

Suppose now that ω is an R-bilinear map: $\Gamma(E) \times \Gamma(E) \to V(M)$ such that:

$$\left.\begin{array}{ll}\text{a)} & \omega(\Psi, \Psi') = - \omega(\Psi, \Psi') \\ & \text{for } \Psi, \Psi' \in \Gamma(E). \\ \text{b)} & \omega\text{, considered as a map of cross-} \\ & \text{sections of vector bundles, is a} \\ & \text{first-order, linear differential} \\ & \text{operator} \\ \text{c)} & \operatorname{div} \omega(\Psi, \Psi') = 0 \text{ if} \\ & 0 = D\Psi = D\Psi'.\end{array}\right. \qquad (3.5)$$

It should be clear now that what we have constructed in special cases in Section 2 is such an

ω satisfying (3.5). In turn, ω defines a skew-symmetric, bilinear, real-valued form on the subspace of $\Gamma(E)$ consisting of the $\Psi \in \Gamma(E)$ such that $D\Psi = 0$ by the rule:

$$\omega_N(\Psi, \Psi') = \int_N \theta_{\omega(\gamma, \gamma')} \qquad (3.6)$$

where N is a fixed submanifold of M whose dimension is one less than the dimension of M.

Let us work out various special cases in terms of local coordinates. Suppose that:

$$E = E' = R^4 \times R^n.$$

As before a point of $M = R^4$ is denoted by $x = (x_\mu)$, $0 \leq \mu, \nu \leq 3$, a point of R^n by $\phi = (\phi_a)$, $1 \leq a, b \leq n$, so that an element of $\Gamma(E)$ is represented by a function $\phi(x) = (\phi_a(x_\mu))$.

Suppose, first, that D is of the following form:

$$(D\phi)_a = \partial_\mu(A_{ab\mu\nu}\partial_\nu\phi_b), \qquad (3.7)$$

where the coefficients $(A_{ab\mu\nu})$ are given functions

of x, such that:

$$A_{ab\mu\nu} = A_{ba\mu\nu}$$
$$A_{ab\nu\mu} = A_{ab\mu\nu}. \tag{3.8}$$

Suppose now that $\phi = (\phi_a(x))$, $\phi' = (\phi'_a)(x))$ are two elements of $\Gamma(E)$. Then,

$$\partial_\mu[A_{ab\mu\nu}(\partial_\nu\phi_b)\phi'_a]$$
$$= (D\phi)_a\phi'_a + A_{ab\mu\nu}\partial_\nu\phi_b\partial_\mu\phi'_a \tag{3.9}$$

$$\partial_\mu(A_{ab\mu\nu}\partial_\nu\phi'_b\phi_a)$$
$$= (D\phi')_a\phi_a + A_{ab\mu\nu}\partial_\nu\phi'_b\partial_\mu\phi_a. \tag{3.10}$$

Set:

$$\omega_\mu(\phi, \phi') = A_{ab\mu\nu}(\partial_\nu\phi_b)\phi'_a$$
$$- A_{ab\mu\nu}(\partial_\nu\phi'_b)\phi_a \tag{3.11}$$
$$\omega(\phi, \phi') = \omega_\mu(\phi, \phi')\partial_\mu.$$

Thus, ω is a first order, bilinear, differential operator: $\Gamma(E) \times \Gamma(E) \to V(M)$, with $M = R^4$.

Condition (3.5a), i.e. that ω is skew-symmetric, is evident from (3.11). Further, we see that as, consequence of (3.8), (3.9) and (3.10), we have:

$$\partial_\mu(\omega_\mu(\phi, \phi')) = (D\phi)_a\phi'_a - (D\phi')_a\phi_a. \quad (3.12)$$

In particular, (3.5c) is satisfied. Thus, the needed "conserved current" can be constructed in case D is of form (3.7).

Now, let us consider the case where the differential operator D is replaced by one of lower order; say, that:

$$(D'\phi)_a = A_{ab\mu}\partial_\mu\phi_b + A_{ab}\phi_b, \quad (3.13)$$

with:

$$\omega'_\mu(\phi, \phi') = B_{ab\mu}\phi_a\phi'_a B_{ab\mu} = - B_{ba\mu}. \quad (3.14)$$

Let us work out the conditions on the coefficient A and B's in (3.13)-(3.14) (which are functions of x) in order that (3.12) be satisfied:

$$(D'\phi)_a\phi'_a - (D'\phi')_a\phi_a$$

$$= (A_{ab\mu}\partial_\mu\phi_b + A_{ab}\phi_b)\phi'_a$$

$$- (A_{ab\mu}\partial_\mu\phi'_b + A_{ab}\phi'_b)\phi_a. \qquad (3.15)$$

$$\partial_\mu\omega'_\mu(\phi, \phi') = \partial_\mu(B_{ab\mu})\phi_a\phi'_b$$
$$+ B_{ab\mu}\partial_\mu\phi_a\phi'_b + B_{ab}\phi_a\partial_\mu\phi'_b. \qquad (3.16)$$

Combining (3.15) and (3.16), we see that we must have:

$$\partial_\mu(B_{ab\mu}) = - A_{ab} + A_{ba} \qquad (3.17)$$

$$B_{ab\mu} = A_{ab\mu} = - A_{ba\mu}. \qquad (3.18)$$

Conversely, conditions (3.17) and (3.18) imply (3.12).

Now, we can combine these two examples:

$$D'' = D + D'.$$

Define:

$$\omega'' = \omega + \omega'.$$

We then see that (3.12) again holds, with D" and

ω'' replacing D and ω. Thus, we have obtained explicit formulas for the "conserved vector currents" associated with a second linear differential operator of the form D". (Higher order operators can be dealt with in a similar way.) In particular, notice that the "linear variational operators" associated with an extremal of the Euler-Lagrange operator derived from a Lagrangian (see Section 2) always satisfies this condition.

4. THE SYMPLECTIC STRUCTURE AND POISSON BRACKET FOR THE HAMILTON EQUATIONS

As we have seen, the space of "extremals" of a variational problem has a "symplectic structure". However, the explicit definition of the symplectic structure is awkward, which creates difficulties when one attempts to calculate the "Poisson bracket" associated with the symplectic structure. In this section, we will outline an alternate approach to the problem using "Hamiltonian" instead of "Lagrangian" ideas.

Introduce the following range of variables and indices:

$0 \leq \mu, \nu \leq 3; \quad 1 \leq i, j \leq 3$

$x = (x_\mu); \; \partial_\mu = \frac{\partial}{\partial x_\mu}$,

$\phi = (\phi_a); \; \phi' = (\phi_{ai})$

$\pi = (\pi_a); \; \pi' = (\pi_{ai})$

$1 \leq a, b \leq n.$

Suppose now that $H(x, \phi, \pi, \phi', \pi')$ is a function of the indicated variables. Introduce the following abbreviation:

$$H_\mu = \frac{\partial H}{\partial x_\mu}$$

$$H_a = \frac{\partial}{\partial \phi_a} \; ; \quad H_{n+a} = \frac{\partial H}{\partial \pi_a}$$

$$H_{ai} = \frac{\partial H}{\partial \phi_{ai}} \; ; \; H_{(n+a)i} = \frac{\partial H}{\partial \pi_{ai}}$$

$$H_{a,b} = \frac{\partial^2 H}{\partial \phi_a \partial \phi_b} \; , \; H_{a,n+b} = \frac{\partial^2 H}{\partial \phi_a \partial \pi_b} \; ,$$

and so forth.

Now, given such a "Hamiltonian" function H, one can define a system of partial differential equations, called the *Hamiltonian equations*, as follows:

$$\begin{aligned}\partial_o\phi_a &= H_{n+a} + \partial_i(H_{(n+a)i})\\ \partial_o\pi_a &= - H_a - \partial_i(H_{ai}).\end{aligned}\tag{4.1}$$

(4.1) is to be "solved" for functions $(\phi_a(x), \pi_a(x))$, $(\phi_{ai}(x), (\pi_{ai}(x))$ are understood to be given in terms of the $(\phi_a(x), \pi_a(x))$ by the following relations:

$$\begin{aligned}\phi_{ai}(x) &= \partial_i\phi_a(x)\\ \pi_{ai}(x) &= \partial_i\pi_a(x).\end{aligned}\tag{4.2}$$

For example, suppose that:

$$H = \frac{1}{2}\,(\pi_a\pi_a + m^2\phi_a\phi_a - \phi_{ai}\phi_{ai}).\tag{4.3}$$

Then, (4.1) and (4.2) together take the following form:

$$\begin{aligned}\partial_o\phi_a &= \pi_a\\ \partial_o\pi_a &= - m^2\phi_a + \partial_i(\partial_i\phi_a(x)).\end{aligned}\tag{4.4}$$

These two first order equations can be combined into the following second order equation:

$$\partial_o{}^2\phi_a - \partial_i\partial_i\phi_a(x) + m^2\phi_a = 0. \tag{4.5}$$

(4.5) is then the "Klein-Gordon equation", that we have treated earlier from the Lagrangian point of view.

Return to the general equations (4.1)-(4.2). Suppose that $f(x, \phi, \pi, \phi', \pi')$ is a real-valued function of the indicated variables. Let M denote the "space" of solutions of (4.1)-(4.2), with a typical element of M denoted by: Ψ: For each real number t, define a real-valued function f on M using the following formula:

$$\underset{\sim}{f}^t(\Psi) = \int f(x, \phi(x), \pi(x), \partial\phi(x), \partial\pi(x)/_{x_o=t}d^3x. \tag{4.6}$$

Thus,

$$\frac{\partial}{\partial t}\underset{\sim}{f}^t(\Psi) = \int(f_o + f_a\partial_o\phi_a + f_{n+a}\partial_o\pi_a + f_{ai}\partial_o\partial_i\phi_a + f_{(n+a)i}\partial_o\partial_i\pi_a)/_{x_o=t}d^3x$$

= , using (4.1),

$$\int (f_o + f_a(H_{n+a} + \partial_i H_{(n+a)i})$$
$$- f_{n+a}(H_a + \partial_i(H_{ai})) + f_{aj}\partial_j(H_{n+a} \qquad (4.7)$$
$$+ \partial_i(H_{(n+a)i}))$$
$$- f_{(n+a)j}\partial_j(H_a + \partial_i(H_{ai}))/_{x_o=t}\, d^3x.$$

Now, let us suppose that the "Cauchy problem" for the system of equations (4.1) is "well-posed", i.e. there is a unique solution of (4.1), reducing to given functions

$$(\phi(0, x_i), \pi(0, x_i))$$

at $t = 0$. Thus, M can be identified with the space of maps: $R^3 \to R^{2n}$, where "R^3" is the space of the x_i, R^{2n} is the space of the variables (ϕ_a, π_a). Alternately, of course, we can construct the fiber space $E = R^3 \times R^{2n}$, and regard M as the space $\Gamma(E)$ of its cross-sections. The space of variables (ϕ, π, ϕ', π') then can be identified with $J^1(E)$, the bundle of one jets. Thus, f or H can be identified with a real-valued function on $J^1(E)$. Now, (4.3) suggests that we define the "Poisson bracket" of f and H, denoted by {f, H}, as the integrand in (4.7). Notice, however, that this is a function of the first and second derivatives of ϕ and π. Thus, the "Poisson bracket" f and H is to be interpreted as a function on $J^2(E)$. We will now follow up this idea, using the ideas of manifold theory. The main point we wish to follow is that E is a fiber space over $N = R^3$, whose fibers have a closed differential 2-form

$$\omega = d\pi_a \wedge d\phi_a$$

which has no characteristic vectors, i.e. defines a "canonical structure" on the fibers. We should then be able to define the "Poisson bracket" more precisely, and elegantly, using the techniques of manifold theory. We will start off with the simplest case - that where the fibers are vector bundles.

5. VECTOR BUNDLES WHOSE FIBERS ARE CANONICAL VECTOR SPACES

First, start off with a definition: Let V be a vector space over the real numbers. A *linear canonical structure* on V is defined by a *skew-symmetric, bilinear* from ω: $V \times V \to R$ which has no non-zero characteristic vectors, i.e.

$$\omega(v, V) = 0 \text{ implies } v = 0. \qquad (5.1)$$

Let $E \to N$ be a vector bundle over a manifold N. Denote a typical point of E by "x", and denote the fiber of E over x by "E(x)". Thus, by the definition of "vector bundle", E(x) is a vector space (say, with the real numbers as scalars). In addition, suppose that we are given, for each $x \in N$, a skew-symmetric, bilinear form ω: $E(x) \times E(x) \to R$, which defines a "linear canonical structure" on E(x), and which varies smoothly with x. This "smoothness" condition can be phrased as follows:

If $\Psi_1, \Psi_2 \in \Gamma(E)$, i.e. are cross-section maps: $N \to E$, then one can define a real-valued function

$$\omega(\Psi_1, \Psi_2): \quad x \to \omega(\Psi_1(x), \Psi_2(x)) \qquad (5.2)$$

on N. We require that this function be "smooth" (or C^∞) on N, i.e. be an element of F(N). Then, (5.2) defines a skew-symmetric, F(N)-linear map,

$$\omega\colon \ \Gamma(E) \times \Gamma(E) \to F(N). \tag{5.3}$$

(Conversely, a "canonical structure" for a vector bundle could be defined as such an algebraic object.)

Now, we must investigate what sort of structure ω induces on the jet bundles. Let us examine this in the special case where N is R^3, i.e. the space of variables $x = (x_i)$, $1 \leq i, j \leq 3$. We will not work completely intrinsically, but will suppose that the F(N)-module $\Gamma(E)$ has a basis, denoted, say, by (Ψ_a), $1 \leq a, b \leq n = (\dim E - \dim N)$. Now, one can show (exercise!) that if an $\Gamma(E)$-basis exists at all, it can be further specialized so that:

$$\omega(\Psi_a, \Psi_b) = \omega_{ab} = \text{constants}. \tag{5.4}$$

Thus, if $\Psi \in \Gamma(E)$, it can be expanded in the form:

$$\Psi = f_a \Psi_a, \tag{5.5}$$

with $f_a \in F(N)$.

Thus, if Ψ, $\Psi' = f'_a\Psi_a \in \Gamma(E)$,

$$\omega(\Psi, \Psi') = f_a f'_b \omega_{ab}. \tag{5.6}$$

Let us work with $J^1(E)$, as an example. Associated with the module basis (Ψ_a), we can define a module basis (Ψ_a, Ψ_{ai}) of $\Gamma(E)$, with the following property:

If $\Psi \in \Gamma(E)$ is of form (5.5), then

$$j^1(\Psi) = f_a\Psi_a + \partial_i(f_a)\Psi_{ai} \tag{5.7}$$

where $\partial_i(f_a)$ denotes the function:

$x \to \frac{\partial}{\partial x_i}(f_a)(x)$.

Of course, $j^1(\Psi)$ - the "one-jet" of the cross-section Ψ, a cross-section of $J^1(E)$, does not exhaust the supply of cross-sections. In general, a cross-section $\Psi^1 \in \Gamma(J^1(E))$ is of the form:

$$\Psi^1 = f_a\Psi_a + f_{ai}\Psi_{ai}, \tag{5.8}$$

with $(f_a, f_{ai}) \varepsilon \Gamma(N)$.

Now, let us define a first-order, linear differential operator

$$D^1: \Gamma(J^1(E)) \rightarrow \Gamma(E) \tag{5.9}$$

as follows:

$$D^1(\Psi^1) = (f_a - \partial_i(f_{ai}))\Psi_a. \tag{5.10}$$

Notice that the definition of D^1 is independent of the form γ. Of course, one can now define a form ω^1 on $\Gamma(J^1(E))$ by the formula:

$$\omega^1(\Psi^1, \Psi^{1'}) = \omega(D^1\Psi^1, D^1\Psi^{1'}) \tag{5.11}$$

for $\Psi^1, \Psi^{1'} \varepsilon \Gamma(J^1(E))$.

Finally, notice that D^1 can be generalized to a map D^r: $\Gamma(J^r(E)) \rightarrow \Gamma(E)$, in a similar way. For example, suppose that $r = 2$. Then, $\Gamma(J^2(E))$ has an F(N)-module basis of the form:

$$(\Psi_a, \Psi_{ai}, \Psi_{aij}), \tag{5.12}$$

with $\Psi_{aij} = \Psi_{aji}$.

Thus, if $\Psi^2 \varepsilon \Gamma(J^2(E))$, it has an expansion of the form:

$$\Psi^2 = f_a \Psi_a + f_{ai} \Psi_i + f_{aij} \Psi_{ij}. \qquad (5.13)$$

Then, define D^2: $\Gamma(J^2(E)) \to \Gamma(E)$, as follows:

$$D^2(\Psi^2) = (f_a - \partial_i(f_{ai}) + \partial_i \partial_j(f_{aij}))\Psi_a. \qquad (5.14)$$

We now turn to showing the connection of all this with quantum field theory.

6. CANONICAL QUANTUM FIELD THEORY, IN THE HEISENBERG PICTURE

Now, we want to show the connection between the material we have been developing on vector bundles and their jets, and quantum field theory. We will work with an "intuitive" picture of quantum field theory, based on the techniques used by elementary particle physicists rather than on any attempt to be rigorous or completely logical and systematic (or "axiomatic"). In a subject as complex (and important) as quantum field theory, with no firmly fixed foundations, but with striking and stimulating links with the physical world, one probably must be content with the traditional (for mathematical physics) sloppy mixture of physical and mathematical intuition, and more or less reasonable, but strictly unjustified assumptions that have always accompanied the development of important mathematical theories of physical phenomena. After all, it took two hundred years to fully explore Newtonian classical mechanics; it is not surprising that in the forty-five years after quantum mechanics everything is not finished. (Of course, these comments are not meant to support the slothful attitude of many physicists towards new mathematics: It seems likely that the ultimate successful development of the theory will call on much of the arsenal of "modern" tools developed by mathematicians, and indeed probably much that is still unknown.)

The ultimate aim of quantum field theory is

to provide "operator-valued functions of space-time points", which will represent the fundamental "fields" needed to describe elementary particles.

Specifically, let us suppose that $(\phi_\alpha(x), \pi_\alpha(x))$ represent such objects. x denotes, in this section, a point of R^4, interpreted as space-time. Thus, $x = (x_\mu)$, $0 \leq \mu, \nu \leq 3$, denotes a point of R^4, with $t = x_0$ the "time" component, (x_i), $1 \leq i, j \leq 3$, the space-components of x. x denotes the 4-vector $(0, x_1, x_2, x_3)$. α, β are "internal symmetry" indices,

$$1 \leq \alpha, \beta \leq n.$$

We will suppose that the $\phi_\alpha(x)$, $\pi_\alpha(x)$ have something like the algebraic properties of skew-Hermitian[1] operators in a Hilbert space. These vectors in the Hilbert space would represent the "states" in the physical system, while the φ's, π's

---

[1]It is customary among physicists to use Hermitian operators both as "observables" and as generators of one-parameter groups of unitary transformations. This places the square root of minus one into awkward places in various commutation relations. Since we will be emphasizing precisely these algebraic aspects of the theory, it seems more convenient to work with skew-Hermitian operators.

and "functions" of them would represent the "observables" of the system. It is typical of the "Heisenberg picture" - as opposed to the "Schrödinger picture" - to emphasize the role of the observables - as detached as possible from their possible role as operators in a Hilbert space - instead of the states. Since it is precisely these algebraic aspects that we mean to emphasize - and this is typical of the insights into quantum field theory obtained from "current algebra" theory - we will follow this idea.

By a "canonical" field theory we mean that the ϕ's and π's satisfy the following "equal time commutation relations":

$$
\begin{aligned}
&[\pi_\alpha(\vec{x}), \phi_\beta(\vec{y})] = \delta_{\alpha\beta}\delta(\vec{x}-\vec{y})I. \\
&0 = [\pi_\alpha(\vec{x}), \pi_\beta(\vec{y})] = [\phi_a(\vec{x}), \phi_\beta(\vec{y})].
\end{aligned}
\tag{6.1}
$$

(I denotes the operator of multiplication of vectors in the Hilbert space by $\sqrt{-1}$.)

One can best interpret this in mathematical terms by introducing "test functions". Let N denote R^3, the space of vectors $\vec{x} = (x_i)$. Let $F_0(N)$

denote the algebra of C^∞, real-valued functions with compact supports.[1] Introduce the formal symbols:

$$\begin{aligned}\phi_\alpha(f) &= \int \phi_\alpha(\vec{x})f(\vec{x})d\vec{x}\\ \pi_\alpha(f) &= \int \pi_\alpha(\vec{x})f(\vec{x})d\vec{x}\end{aligned} \tag{6.2}$$

for $f \in F_0(N)$.

Then, using (6.1) in the usual formal way for dealing with the "Dirac delta functions", gives the relations:

$$\begin{aligned} 0 = [\phi_\alpha(f_1), \phi_\beta(f_2)] &= [\pi_\alpha(f_1), \pi_\beta(f_2)]\\ [\pi_\alpha(f_1), \phi_\beta(f_2)] &= I\delta_{\alpha\beta}(\int f_1 f_2(x)dx)\end{aligned} \tag{6.3}$$

for $f_1, f_2 \in F_0(N)$.

We can interpret this as follows: Introduce Γ as the real vector space generated by the "symbols" $(\pi_\alpha(f), \phi_\alpha(f))$, where f runs over all of

[1] This means that, for each $f \in F_0(N)$, the function $\vec{x} \in f(\vec{x})$ vanishes outside of *some* bounded subset of R^3. (The subset may vary with f.)

F(N). Then, Γ can be designed as an F(N)-module in the following way:

$$f_1(\pi_\alpha(f)) = \pi_\alpha(f_1 f)$$

$$f_1(\phi_\alpha(f)) = \phi_\alpha(f_1 f) \tag{6.4}$$

for $f_1 \in F(N)$, $f \in F(N)$.

Then, Γ so defined is a *free* F(N)-module; it has a basis consisting of the elements $(\pi_\alpha(1), \phi_\alpha(1))$, which generate a real vector space V of dimension 2n. Thus, Γ can be considered in a natural way as the space Γ(E) of cross-sections of the vector bundle $E = N \times V$.

This remark leads to a very natural mathematical interpretation of (6.4). Define a skew-symmetric, bilinear form ω, i.e. a "linear canonical structure" on V, as follows:

$$\omega(\pi_\alpha(1), \phi_\beta(1)) = \delta_{\alpha\beta}.$$

$$\omega(\pi_\alpha(1), \pi_\beta(1)) = 0 = \omega(\phi_\alpha(1), \phi_\beta(1)). \tag{6.5}$$

Extend ω to be an F(N)-bilinear map

$$: \Gamma \times \Gamma \to F(N),$$

as explained in Section 5. Then, define a skew-symmetric *real* bilinear form on Γ, as follows:

$$(\Psi_1, \Psi_2) \to \int \omega(\Psi_1, \Psi_2)(\vec{x})d\vec{x}. \qquad (6.6)$$

The direct sum of Γ with an element "1", defines a vector space, which we will call "$\underset{\sim}{H}$". $\underset{\sim}{H}$ can be made into a real Lie algebra, as follows:

$$\begin{aligned} &[1, \Psi] = 0 \\ &[\Psi_1, \Psi_2] = (\int\omega(\Psi_1, \Psi_2)(\vec{x})d\vec{x}) \end{aligned} \qquad (6.7)$$

for $\Psi_1, \Psi_2 \in \Gamma$.

This Lie algebra - *the field-theoretic Heisenberg algebra* - plays a role in quantum field theory analogous to the Lie algebra of the p's and q's in particle quantum mechanics. Notice that the *operator* commutation relations (6.3) are just a homomorphic image of the "abstract" Lie algebra commutation relations (6.6). This suggests that we study H independently of its realization as operators, and seek to express physical ideas directly in terms of the Lie algebra and geometric structure[1] of H. Translated into our language, this is what is meant by the "Heisenberg picture".

Let us return to the "primitive" idea of the $(\pi_\alpha(x), \phi_\beta(x))$ as operator-valued "functions" of space-time points x. We can then define the "derivatives"

$$\partial_\mu \pi_\alpha, \ \partial_\mu \phi_\beta, \ \partial_\mu \pi_\alpha, \ \text{etc.}$$

of these objects in the usual way. $\left(\partial_\mu = \frac{\partial}{\partial x_\mu}\right)$. We must now show how these derivatives can be interpreted in the "Heisenberg picture" language.

Unfortunately, it is necessary to destroy "manifest" Lorentz symmetry, and treat the space and time derivatives differently. First, we will deal with the space derivatives ∂_i, $1 \leq i, j \leq 3$. Set:

$$\begin{aligned} \pi_{ai} &= \partial_i \pi_\alpha, \ \phi_{ai} = \partial_i \phi_a \\ \pi_{aij} &= \partial_i \partial_j \pi_\alpha, \quad \text{and so forth.} \end{aligned} \tag{6.8}$$

[1]By "geometric structure" we mean its relation to the vector bundle structures over R^3 and R^4.

Let $E = N \times V$, with $N = R^3$, be the vector bundle described above, associated with the "fields" $\pi_\alpha(\vec{x})$, $\phi_\alpha(\vec{x})$ at zero time. Let $J^1(E)$ be the jet bundle of its cross-sections. Let $\Gamma_0(E)$ and $\Gamma_0(J^1(E))$ be the F(N)-module consisting of its compact support cross-sections. As we have seen, a $\Psi \in \Gamma_0(E)$ can be written in the following form:

$$\Psi = f_\alpha \pi_\alpha(1) + f'_\alpha \phi_\alpha(1), \tag{6.9}$$

with f_α, $f'_\alpha \in \Gamma_0(N)$.

As an F(N)-module, $\Gamma(J^1(E))$ has a basis consisting of the elements $(\pi_\alpha(1), \phi_\alpha(1))$, together with additional elements

$$\pi_{\alpha i}, \phi_{\alpha i}.$$

Thus, a $\Psi^1 \in \Gamma_0(J^1(E))$ can be written as:

$$\Psi^1 = f_\alpha \pi_\alpha(1) + f'_\alpha \phi_\alpha(1) + f_{\alpha i}\pi_{\alpha i} + f'_{\alpha i}\phi_{\alpha i}. \tag{6.10}$$

Let us associate with Ψ^1 of form (6.10) the following "operator"

$$\int[f_\alpha(\vec{x})\pi_\alpha(1) + f'_\alpha(\vec{x})\phi_\alpha(1) + f_{\alpha i}(\vec{x})\partial_i\pi_\alpha(\vec{x}) + f'_{\alpha i}(\vec{x})\partial_i\phi_\alpha(\vec{x})]dx. \tag{6.11}$$

Since the coefficient f's are of compact support, the last two terms in (6.10) can be integrated by parts, leading to the expression:

$$\int[\pi_\alpha(f_\alpha)(\vec{x}) + \phi_\alpha(f_\alpha)(\vec{x}) - \pi_\alpha(\partial_i f_{\alpha i})(\vec{x}) - \phi_\alpha(\partial_i f'_{\alpha i})(\vec{x})]dx. \tag{6.12}$$

(6.12) then tells us that the assignment - via (6.11) - of an "operator" to the element $\Psi^1 \in \Gamma(J^1(E))$ can be accomplished by assigning the element $D^1(\Psi^1) \in \Gamma_0(E)$ to Ψ^1, then following the rules (6.2), where $D^1\colon\ \Gamma(J^1(E)) \to \Gamma(E)$ is the first-order linear differential operator constructed in Section 5. Thus, we see a way of introducing abstractly "commutation relations" for the (space) derivatives of quantum fields.

How to handle the "time" derivatives ∂_0 is not quite so obvious. We will now describe the most naive way, as a first approximation.

If $(\phi_\alpha(x), \pi_\alpha(x))$ are "quantum fields", depending on a space-time vector $x = (x_\mu)$, introduce objects labelled

$$(\phi_\alpha{}^t(f), \pi_\alpha{}^t(f)), \ t \in R, \ f \in F_0(N)),$$

as follows:

$$\begin{aligned} \phi_\alpha{}^t(f) &= \int f(\vec{x})\phi_\alpha(t, \vec{x})d\vec{x} \\ \pi_\alpha{}^t(f) &= \int f(\vec{x})\pi_\alpha(t, \vec{x})d\vec{x}. \end{aligned} \tag{6.13}$$

In this picture, we regard the $\phi_\alpha{}^t(f)$, $\pi_\alpha{}^t(f)$ as "curves" (with "t" the "curve parameter") in the Lie algebra of operators. Then in the time-derivatives

$$\frac{\partial}{\partial t}\phi_\alpha{}^t(f), \ \frac{\partial}{\partial t}\pi_\alpha{}^t(f)$$

are defined as usual. Formally, of course, we have:

$$\begin{aligned} \frac{\partial}{\partial t}\phi_\alpha{}^t(f) &= \int f(\vec{x})\partial_0\phi_\alpha(t, \vec{x})d\vec{x} \\ \frac{\partial}{\partial t}\pi_\alpha{}^t(f) &= \int f(\vec{x})\partial_0\pi_\alpha(t, \vec{x})d\vec{x}. \end{aligned} \tag{6.14}$$

Formula (6.14) enables us to translate back and forth between statements in terms of the "manifestly Lorentz covariant" objects $\partial_\mu \phi_\alpha$, $\partial_\mu \pi_\alpha$, and the Heisenberg picture objects.

Now, the main question of "dynamics" in quantum field theory is the description of these time derivatives in terms of the fields themselves. The most naive approach proceeds as follows:

Regard the (ϕ_α) as coordinates of a vector space V', and construct the vector bundle

$$E' = R^4 \times V'.$$

Let $L(\phi, \partial\phi)$ be a "Lagrangian" for this vector bundle. Using L, construct the Euler-Lagrange equations:

$$\partial_\mu (L_{\alpha\mu}(\phi(x), \partial\phi(x))) = L_\alpha(\phi(x), \partial\phi(x)). \quad (6.15)$$

Now, in the usual interpretation of (6.15), the $\phi(x)$ represents "classical fields", i.e. real-valued functions of space-time points. One now attempts to make sense of (6.15) as a differential equation between *operator valued* functions of x, with $\pi_\alpha(x)$ defined - in "canonical" field theories - by the relation:

$$\pi_\alpha(x) = L_{\alpha 0}(\phi(x), \partial\phi(x)). \tag{6.16}$$

In general, this program meets with great difficulties, although no doubt the work that has been done (most successfully, in the quantum electrodynamics case) will serve as a stepping stone to the ultimate theory.

Now, one immediate and obvious source of difficulty in the interpretation on the quantum level of the "field equations" (6.15) is that the $\phi_\alpha(x)$ are to be operators, that do not commute in general. Thus, there will be difficulties in the interpretation of what is meant in (6.1) by non-linear "functions" of the $\phi_\alpha(x)$. Of course, if the equations (6.15) are linear, this particular difficulty will not arise. The fields satisfying such a linear equation are called *free fields*. There is available a very rigorous theory of such "free fields". (See Streater-Wightman [1] and Jost [1].) One can regard these "free fields" as a mathematical testing ground for attempts to construct a general mathematical theory of quantized fields.

For example, the simplest case is that where the $\phi_\alpha(x)$ satisfying the Klein-Gordon equation:

$$\partial_0 \partial_0 \phi_\alpha - \partial_i \partial_i \phi_\alpha = m^2 \phi_\alpha. \tag{6.17}$$

In this case,

$$\pi_\alpha = \partial_0 \phi_\alpha, \tag{6.18}$$

and the system $(\phi_\alpha, \pi_\alpha)$ satisfies a system of first order linear differential equations:

$$\begin{aligned} \partial_0 \phi_\alpha &= \pi_\alpha \\ \partial_0 \pi_\alpha &= \Delta \phi_\alpha - m^2 \phi_\alpha, \end{aligned} \tag{6.19}$$

where $\Delta = \partial_i \partial_i$ as the Laplacian operator. These equations can then be translated into the "Heisenberg picture", as follows; using (6.14)

$$\begin{aligned} \frac{\partial}{\partial t} \phi_\alpha{}^t(f) &= \pi_\alpha{}^t(f) \\ \frac{\partial}{\partial t} \pi_\alpha{}^t(f) &= \phi_\alpha(\Delta(f)) - m^2 \phi_\alpha(f) \end{aligned} \tag{6.20}$$

for $\quad f \in \Gamma(N)$.

The "solution" of (6.15) in terms of the "Cauchy data" $\phi_\alpha{}^0(f)$, $\pi_\alpha{}^0(f)$ can now be readily written down in terms of a "Green's function" (or "fundamental solution") for the Laplace-Helmholtz differential operator: $\Delta - m^2$: In turn, notice that the right hand side of (6.15) is a vector bundle linear differential operator: $\Gamma(E) \to \Gamma(E)$. The "Green's function" is essentially a pseudo-differential operator (see Hormander [1]) on this vector bundle. We see immediately that there will be an intimate connection between the theory of differential and pseudo-differential operators on vector bundles and, at least, the theory of free quantum fields. We will now pause in our exposition of the mathematical ideas of quantum field theory in order to develop subsidiary algebraic and group-theoretic ideas.

CHAPTER VI

VECTOR BUNDLES AND REPRESENTATIONS OF SEMIDIRECT PRODUCT GROUPS

Our aim is to prepare the way for the study (in Chapter 7) of differential operators on vector bundles over R^4 that are invariant under the action of the Lorentz and Poincaré group. The development of some general principles is useful for this end, and also proves useful as a unifying force in several seemingly distinct areas of mathematics and physics. Thus, we will pause in our development of the main theme of this book - the connection between quantum field theory and the theory of vector bundles - in order to explain some of these unifying threads.

1. VECTOR BUNDLES ARISING BY MEANS OF SEMIDIRECT PRODUCT GROUP REPRESENTATIONS

In this chapter, it will be most convenient to suppose that all vector spaces are - unless mentioned otherwise - defined with the complex numbers as their field of scalars.

Let G be a group, and let H be a vector space. A *(linear) representation of* G *on* H is defined as a homomorphism - typically denoted by ρ - of G into the group of invertible linear transformations on H. In other words, G acts as a transformation group on H - with the transform of $\Psi \in H$ by $g \in G$ denoted by $\rho(g)(\Psi)$ - and with each $\rho(g)$ a linear transformation of H. If G is a "topological vector space", it is customary to assume also that all the data at least is "continuous". However, we will not bother to keep track of this type of assumption, but will emphasize other details that seem more important for the applications.

If H and H' are vector spaces, if ρ, ρ' are

representations of groups G and G' on H and H', and if ϕ: $G \to G'$ is a group homomorphism, a linear map α: $H \to H'$ is an *intertwining map* if:

$$\alpha(\rho(g)(\Psi)) = \rho'(\phi(g))(\alpha(\Psi))$$

$$\text{for} \quad g \in G, \ \Psi \in H.$$

Very often, G will equal G' and ϕ is the identity map. If α is an invertible map, i.e. if α^{-1}: $H' \to H$ exists, then the representations ρ and ρ' are said to be *equivalent*. A basic mathematical problem is to enumerate the equivalence classes of representations of a given type of groups, and provide a useful method of construction of typical elements of each equivalence class. For example, if one restricts attention to finite dimensional representations of semisimple, compact Lie groups G, this problem was solved by Frobenius, Schur, E. Cartan and H. Weyl in their classic work. We will now discuss this basic problem in case G is a "semidirect product group", with the invariant subgroup the abelian case. Here, the methods were developed by Frobenius, Schur, Wigner, and Mackey.

First, we will briefly explain some of the

group-theoretic terminology we will be using. Let G be a group. The product of two elements g_1, g_2 (following the group-law defining G) is denoted by $g_1 g_2$.

A subset K of G is a *subgroup* if:

$$g_1 g_2 \in K \quad \text{for} \quad g_1, g_2 \in K$$

$$g^{-1} \in K \quad \text{for} \quad g \in K.$$

A subgroup K is an *invariant subgroup* if:

$$gkg^{-1} \in K, \quad \text{for} \quad g \in G, k \in K.$$

A group G is a *semidirect product* of a subgroup S and an invariant subgroup K if:

$$K \cap S = (e) \tag{1.1}$$

(e denotes the identity element of the group)

$$G = KS, \tag{1.2}$$

i.e. each element of G can be written in at least one way as the product of an element of K and an element of S. In fact, this representation is unique, as the following result shows:

<u>LEMMA 1.1</u>. If (1.1) and (1.2) holds, and if g is an element of G such that:

$$g = sk = s'k', \quad \text{with} \quad s, s' \in S, \ k, k' \in K,$$

then: $s = s'$, $k = k'$.

<u>Proof</u>: We have:

$$kk'^{-1} = s^{-1}s'.$$

By (1.1), both sides of this relation must be the identity element of G.

Conversely, given groups K and S, we may ask how semidirect product groups G may be constructed containing them. To answer this question, notice that, if such a G exists, then for each $s \in S$, there is an isomorphism ϕ_s of K into itself, given by:

$$\phi_s(k) = sks^{-1} \quad \text{for} \quad k \in K.$$

Let Iso (K) denote the group of isomorphisms of K into itself (with the group product just map-composition). Thus, $s \to \phi_s$ determines a homomorphism of $S \to \text{Iso}(K)$.

Now, suppose G is not assumed to exist, but a homomorphism ϕ: $S \to \text{Iso}(K)$ is given. We can construct a G that is a semidirect product of S and K, in the following way:

> G, as a set of points, is $S \times K$, the Cartesian product of S and K. However, the group law is defined in the following way:

$$(s, k)(s', k') = (ss', \phi_{s'^{-1}}(k)k'). \qquad (1.3)$$

Notice that (1.3) represents a "twisted" version of the "direct product" law:

$$(s, k)(s', k') = (ss', kk'). \qquad (1.4)$$

Of course, (1.3) reduces to (1.4) if ϕ_s = identity.

It is often convenient to use the notation: $G = S \cdot K$: for the semidirect product. If the homomorphism ϕ that determines G needs to be mentioned explicitly, the following notation may be used: $G = S\cdot_\phi K$:

With these notations in hand, let us turn to the study of the representations of G. Let $\rho(G)$ be a linear representation of G on a vector space H. We will not study the most general possible representation, but will make certain assumptions (that are usually very natural in the examples from physics).

Recall that an *eigenvector* of $\rho(K)$ is a non-zero element $\Psi \in H$ such that:

$$\rho(k)(\Psi) = \lambda(k)\Psi \tag{1.5}$$

for all $k \in K$.

The complex number $\lambda(k)$ appearing on the right hand side of (1.5) - that is, the *eigenvalue* - depends, of course, on Ψ and k. As a function of k, it is a mapping: $K \to C$ (C = complex numbers) such that:

$$\lambda(k_1 k_2) = \lambda(k_1)\lambda(k_2) \tag{1.6}$$

for $k_1, k_2 \in K$.

Now, any mapping λ: $K \to C$ satisfying (1.6) - that is, a homomorphism between K and the group (under multiplication) of non-zero complex numbers - is called a *character* of K. The set of such characters is denoted by: K^d: (read "the dual group"). K^d is also a group: The product of $\lambda_1, \lambda_2 \in K^d$ is defined as λ: $k \to \lambda(k) = \lambda_1(k)\lambda_2(k)$.

Now, G acts in a natural way as a group of automorphisms of K^d:

$$(g\lambda)(k) = \lambda(g^{-1}kg) \qquad (1.7)$$
$$\text{for} \quad g \in G,\ k \in K,\ \lambda \in K^d.$$

Now, given $\lambda \in K^d$, define a linear subspace H^λ of H as follows:

H^λ = space of vectors $\Psi \in H$ such that

$$\rho(k)(\Psi) = \lambda(k)\Psi \qquad (1.8)$$

for all $k \in K$.

Notice the following rule.

$$\rho(g)(H^\lambda) = H^{g\lambda}, \qquad (1.9)$$

where $g\lambda$ is defined by (1.7).

<u>Proof of (1.9)</u>. If $\Psi \in H^\lambda$,

$$\rho(k)(\rho(g)(\Psi))$$
$$= \rho(g)\rho(g^{-1}kg)(\Psi)$$
$$= \rho(g)(\lambda(g^{-1}kg)(\Psi))$$
$$= \lambda(g^{-1}kg)\rho(g)(\Psi)$$
$$= (g\lambda)(k)(\rho(g)(\Psi)).$$

This shows that $\rho(g)(\Psi) \in H^{g\lambda}$, hence proves (1.9).

Now, define M as the subspace of K^d consisting of the $\lambda \in K^d$ for which $H^\lambda \neq (0)$, i.e. M^ρ is the collection of "eigenvalues" of $\rho(K)$. Define a vector bundle E, whose base space is M^ρ, as follows:

> E is the subspace of $K^d \times H$, consisting of the points $(\lambda, \Psi) \in K^d \times H$ such that:
>
> a) $\lambda \in M^\rho$.
>
> b) $\Psi \in H^\lambda$. (1.10)

The projection π: $E \to M^\rho$ is defined as follows:

$$\pi(\lambda, \Psi) = \lambda.$$

Thus, the fiber $E(\lambda) = \pi^{-1}(\lambda)$, for $\lambda \in M^\rho$, consists of H^λ. It is a vector space, in a natural way, since it is a linear subspace of H. This, then, defines (E, M^ρ, π) as a "vector bundle", at least in a primitive form. (It is not necessarily a "local product" bundle.) It has at least enough features of a "vector bundle" that we can attempt to use the geometric intuition developed rigorously for "honest" vector bundles.

What is the relation between the vector bundle E and the vector space H with which we started? First of all, we can construct a linear map

$$\alpha: \ \Gamma(E) \to H$$

in the following way.

Let us suppose, for example, that M^ρ is a manifold with a volume element form "$d\lambda$". Then, we can define α by the formula:

$$\alpha(\Psi') = \int_{M^\rho} \Psi(\lambda)d\lambda \qquad (1.11)$$

for $\Psi \in \Gamma(E)$.

Of course, (1.11) is somewhat "formal"; to make it precise we would have to specify what it means to "integrate" a function on M^ρ taking values in H. Now, H may be an infinite dimensional vector space - indeed, this is the most interesting case. However, we will not pursue these technicalities here, at least for the moment.

Now we ask: Can we find a group action ρ' of G on $\Gamma(E)$ so that the map (1.11) intertwines? To answer this, let us look for ρ' of the form:

$$\rho'(g)(\Psi')(\lambda) = m(g, \lambda)\rho(g)(\Psi'(g^{-1}\lambda)), \quad (1.12)$$

$$\text{for} \quad g \in G, \Psi' \in \Gamma(E), \lambda \in M^\rho,$$

where $m(g, \lambda)$ is a complex-valued function of the indicated variables satisfying the group-representation "multiplier" functional equation (see "Lie groups for physicists"), i.e. just that equation needed to assure the fact that (1.12) defines a genuine representation of G.

Let us look at the conditions m has to satisfy in order that α intertwines ρ' and ρ. Now, for $\Psi' \in \Gamma(E)$,

$$\alpha(\rho'(g)(\Psi')) = \int m(g, \lambda)\rho(g)(\Psi'(g^{-1}\lambda))d\lambda \tag{1.13}$$

Let $j_g(\lambda)$ be the "Jacobian" of the transformation $\lambda \to g\lambda$ of M^ρ with respect to the volume element form "$d\lambda$", i.e.

$$\int_{M^\rho} f(g^{-1}\lambda)d\lambda = \int f(\lambda)j_g(\lambda)d\lambda \tag{1.14}$$

for $f \in F(M^\rho)$.

Thus, the right hand side of (1.12) can be rewritten as:

$$\int m(g, g\lambda)j_g(\lambda)\rho(g)(\Psi(\lambda))d\lambda. \tag{1.15}$$

Suppose now that:

$m(g, g\lambda)j_g(\lambda)$ is independent of λ, (1.16)

say, equal to $h(g)$.

Then, (1.15) can be rewritten

$$h(g)\rho(g)\alpha(\Psi'). \tag{1.17}$$

We can conclude that if:

$$h(g) = 1, \tag{1.18}$$

then α is indeed an intertwining operator. Condition (1.18), in turn, determines $m(g, \lambda)$, since we then have:

$$m(g, \lambda) = j_g(g^{-1}\lambda)^{-1}. \tag{1.19}$$

Conversely, one proves readily that if (1.19) defines m, it satisfies the multiplier operator, hence ρ' is a genuine representation.

In many of the important cases - for example, the case where G is the Poincaré group - j_g is identically one, i.e. the measure can be chosen on M^ρ which is strictly invariant under the action of G.

We will not pursue further the study of this map α as proper mathematical object in the general case where H is infinite dimensional and G is a general Lie group. To give some idea what to expect in these general cases, we will consider the following simple-minded result:

THEOREM 1.1. Suppose that G is a finite group, with K abelian, that H is a finite dimensional vector space, and that $\rho(G)$ is an irreducible set of operators. Then, the following conditions are satisfied:

a) α is an isomorphism between the vector spaces $\Gamma(E)$ and H

b) G acts transitively on M^{ρ}. (1.20)

c) Let λ_0 be a fixed point of M^{ρ}, and let L be the isotropy subgroup of G at λ_0. Then, $\rho(L)$ - which is just the linear isotropy representation of L on $E(\lambda_0)$ - acts irreducibly in $H^{\lambda_0} = E(\lambda_0)$.

Conversely, if conditions (1.20) are satisfied, then $\rho'(G)$ defines an irreducible representation of G by operators in $\Gamma(E)$.

Proof. Since G is supposed finite and H is finite dimensional, there is no loss in generality in supposing that H is a Hilbert space, and that $\rho(G)$ are unitary operators. For Ψ_1, $\Psi_2 \in H$, let $\langle\Psi_1/\Psi_2\rangle$ be the Hilbert-space inner product between these two vectors.

Then, a well-known property of eigenvectors of unitary operators is that they are orthogonal if they correspond to different eigenvalues. Thus,

$$<H^{\lambda_1}/H^{\lambda_2}> = 0 \tag{1.21}$$

if $\lambda_1, \lambda_2 \in M^\rho$, $\lambda_1 \neq \lambda_2$.

Since the operators $\rho(K)$ are abelian and unitary, they can be brought to simultaneous matrix diagonal form; this means that H is the direct sum of the subspaces H^λ, where λ runs through M^ρ. Now, M^ρ is a finite set of points. Choose the volume element "$d\lambda$" so that it assigns equal weight to each point of M^ρ. Then, we see the statement:

$$H = H^{\lambda_1} \oplus H^{\lambda_2} \oplus \ldots \oplus H^{\lambda_n}$$

where $\lambda_1, \ldots, \lambda_n$ are the points of M^ρ,

is equivalent to the statement that α defined by (1.11) is an isomorphism. (Of course, the great virtues of the admittedly more complicated method of formulation in (1.11) is that it suggests generalizations to more complicated situations, and carries with it the geometric intuition associated with the theory of vector bundles).

Suppose now that N is a subset of M^{ρ} that is invariant under the action of G. One can construct a vector bundle E' on N by assigning to each point $\lambda \in N$ the fiber $E(\lambda)$ of E over λ (E' is called the *subbundle of* E *obtained by restricting* E to N.) Then, using a formula similar to (1.11), one can construct an intertwining map $\alpha'\colon\ \Gamma(E') \to H$. Then,

$$\rho(G)(\alpha'(\Gamma(E')) \subset \alpha'(\Gamma(E')).$$

In particular, if $\rho(G)$ acts irreducibly,

$$\alpha'(\Gamma(E')) = H. \tag{1.22}$$

Now, $\dim \Gamma(E') \leq \dim \Gamma(E) = \dim H.$ (1.23)

(1.22) and (1.23) force the relation:

$$\dim \Gamma(E) = \dim \Gamma(E'). \tag{1.24}$$

However, (1.24) is only possible if:

$$N = M^{\rho},$$

i.e. G acts transitively on M^{ρ}. This proves (1.20b).

Turn to the proof of (1.20c). Suppose that V is a subspace of $H^{\lambda_0} = E(\lambda_0)$ that is invariant under the action of $\rho(L)$. One can define a vector bundle E' over M^{ρ} as follows:

> E" is the subset of E consisting of the vectors v which can be obtained by translating under $\rho(G)$ a vector of V.

(In other words, fiber of E" over a point $\lambda = g\lambda_0$ of M^{ρ} is the vector space $\rho(g)(V)$). Again, E" is a subbundle of E which is invariant under the action of G. Again, we can construct an intertwining map $\alpha''\Gamma(E'') \to H$. Again, irreducibility of $\rho(G)$ forces:

$$\dim \Gamma(E'') = \dim \Gamma(E),$$

hence: $E = E''$, i.e.

$$V = E(\lambda_0) = H^{\lambda_0}.$$

Then, $\rho(L)$ must act arreducibly on $E(\lambda_0) = H^{\lambda_0}$, which proves c).

The proof of the converse involves the study of "intertwining operators", a topic we will put off into Volume 3 of this work.

Notice that all these arguments have a broad, "geometric" flavor that transcends the special case of "unitary representations" and that will be useful later on. Let $G = S \cdot K$ be the semidirect product of a group S and an abelian invariant subgroup K, with $\rho(G)$ a representation of G by operators on a complex vector space H. Let us suppose further that H has a "Hilbert space structure" with respect to which $\rho(G)$ are "unitary operators". However, we do not necessarily mean these terms to have the precise meaning now standard in the mathematics literature (but a meaning closer to the usage in the physics literature). Instead, we only require the following sort of structures:

a) H has a real-bilinear, complex-valued inner product, $(\Psi_1, \Psi_2) \to \langle\Psi_1/\Psi_2\rangle$, such that:

$$<\Psi_1/\Psi_2> = <\Psi_2/\Psi_1>^*$$

$$<c\Psi_1/\Psi_2> = c^*<\Psi_1/\Psi_2> = <\Psi_1/c^*\Psi_2>$$

$$<\Psi_1/\Psi_1> > 0$$

for $c \in C$, Ψ_1, $\Psi_2 \in H$. (* denotes "complex conjugate). These properties can be called a "Hermitian-symmetric, positive, inner product."

b) For $g \in G$,

$$<\rho(g)\Psi_1/\rho(g)\Psi_2> = <\Psi_1/\Psi_2>$$

for Ψ_1, $\Psi_2 \in H$.

Then, if $\Psi \in H$ is an eigenvector of the abelian operators $\rho(K)$, one expects that:

$$\rho(k)(\Psi) = \lambda(k)\Psi, \tag{1.25}$$

where $k \to \lambda(k)$ is a map: $K \to$ (complex numbers of absolute value one) = $U(1)$. Notice that the image group $U(1)$ is an abelian group under multiplication. Thus, the set of homomorphisms λ from K to this group forms a group itself, since two such homomorphisms can be multiplied to obtain a third one. This group may be called the *unitary dual group* of K. We will not use our special notation for this group, but will denote if also by "K^d", depending on the reader to sort out from the context which meaning is assigned to K^d. (In fact, in the sequal, the "unitary" dual group will be used most often and is most immediately relevant to the applications we have in mind. Thus, the reader, to the first approximation, may assume that our discussion deals always with "unitary representations" and "unitary" dual groups, the vector bundles constructed from the unitary dual groups, etc. We apologize in advance to the reader, for any confusions he might encounter, but it is not really worthwhile burdening our notations with the additional apparatus needed to be precise on this point.)

2. HOMOGENEOUS VECTOR BUNDLES AND INDUCED REPRESENTATIONS

Let π: $E \to M$ be a vector bundle over a manifold M, with a Lie group G acting on E and M, and acting linearly on E. Suppose G acts transitively on M. Then, this set-up will be called a *homogeneous vector bundle.* (The "homogeneity" refers to the fact that fibers above points of M can be transported from one point to the other by the action of G.)

Now, E can be reconstructed from $\Gamma(E)$. In fact, given $p \in M$, $E(p)$ is just the quotient of $\Gamma(E)$ by the subspace of those $\Psi \in \Gamma(E)$ that vanish at p. In turn $\Gamma(E)$ can be described in terms of vector-valued functions on G. (This connection is described in more detail in "Lie groups for physicists".)

In fact, pick a fixed point $p_0 \in M$. Let L be the isotropy of $g \in G$ such that: $gp_0 = p$: Let V be the fiber $E(p_0) = \pi^{-1}(p_0)$ of E at p_0. Then, M can be identified with the coset space G/L.

For each $\Psi \in \Gamma(E)$, let the map f_Ψ: $G \to V$ be constructed as follows:

$$f_\Psi(g) = g^{-1}\Psi(gp_0)$$

for $g \in G$.

Let $\sigma(L)$ be the representation of L by operators on $V = E(p_0)$, the "linear isotropy representation" of the vector bundle. Then,

$$f_\Psi(g\ell) = \ell^{-1}g^{-1}\Psi(g\ell p_0)$$

$$= \ell^{-1}g^{-1}\Psi(gp_0)$$

$$= \sigma(\ell^{-1})(f_\Psi(g)).$$

Thus, f_Ψ belongs to the class of maps $f\colon\ G \to V$ satisfying the following transformation law under right translation by L:

$$f(g\ell) = \sigma(\ell^{-1})f(g)$$

$$\text{for} \quad g \in G,\ \ell \in L.$$

It can be shown (exercise) that this map $\Psi \to f_\Psi$ is an isomorphism between $\Gamma(E)$ and the vector space of functions $f\colon\ G \to V$ satisfying (2.1). To make this map an intertwining map relative to the action of G on $\Gamma(E)$, one has only to define the action of $g \in G$ on f as follows:

$$(g_0f)(g) = f(g_0^{-1}g) \tag{2.2}$$

$$\text{for} \quad g \in G.$$

The representations given by (2.2) of G on the vector space of the f's defined by (2.1) is called the *representation of* G *induced from the representation* σ *of the subgroup L*. In view of these remarks, it can just as well be defined as the action of G on the cross-sections Γ(E).

As a bonus from these arguments, we see that the homogeneous vector bundle is determined up to isomorphism by the representation σ of L, and conversely to each such representation σ there is at least one vector bundle for which it is the linear isotropy representation. For, as we have remarked above, E can be reconstructed from Γ(E), and Γ(E) can be constructed - as the space of f's - given σ.

Now, we transform these generalities to some situations of interest in physics.

3. THE EUCLIDEAN MOTION GROUP

First, we will indicate briefly what the "Euclidean motion group" is.

Consider R^3, "our" 3-dimensional, Euclidean space. Denote typical points of R^3 by x, y,..., with $x = (x_i)$, $y = (y_i)$, $1 \leq i, j \leq 3$. The "inner product" and "distance" functions are defined as follows:

$$x \cdot y = x_i y_i$$

$$|x-y| = \sqrt{(x-y)\cdot(x-y)}.$$

Let E(3) = G be the group of transformations: $R^3 \to R^3$ which preserve distances between points. (G is called the *Euclidean motion group*, whence the notation "E(3)".) It has two immediately interesting subgroups:

a) Let K = O(3, R) be the group of distance preserving *linear* transformations: $R^3 \to R^3$. One shows readily that $k \in K$ is an *orthogonal transformation* (whence the notation "O(3, R)"), in the sense that:

$$(kx)\cdot(ky) = x \cdot y$$

$$\text{for} \quad x, y \in R^3$$

b) Let T be the group of translations. T is isomorphic to the additive group of the vector space R^3. For each $a \in R^3$, $g_a = R^3 \to R^3$ is defined as follows:

$$g(x) = x + a.$$

Now, one sees that G is the semidirect product of K and T. For, given $g \in G$, set: $a = -g(0)$. Then,

$$g_a g(0) = 0.$$

Hence, $g_a g$ is a distance preserving map: $R^3 \to R^3$ which maps zero into zero. One proves then that $g_0 g$ must belong to K (exercise).

Hence, $g \in TK$, i.e. $G = TK$. Further,

$$T \cap K = (\text{identity}).$$

T is an invariant subgroup. (3.1)

For, given $k \in K$, $g_a \in T$, $x \in R^3$

$$(kg_a k^{-1})(x) = k(k^{-1}x + a) = x + k(a),$$

i.e.

$$kg_a k^{-1} = g_{k(a)}. \qquad (3.2)$$

This not only proves (3.1) - hence completes the proof that G is a semidirect product K•T - but shows that the homomorphism: $K \to$ (group of isomorphisms of $T = R^3$) defining the semidirect product (see Section 1) is essentially the given linear action of $K = O(3, R)$ on R^3.

Now, the group G is the basic "symmetry group" of a free particle in non-relativistic, Schrödinger-quantum mechanics. We can explain what is meant here in the following way.

Let H be the vector space of C^∞, complex-valued functions $x \to \Psi(x)$ defined on R^3. Let E be the following linear operator: $H \to H$:

$$E(\Psi) = \frac{1}{2} \Delta(\Psi) \tag{3.3}$$

where $\Delta = \partial_i \partial_i$.

$$\left(\partial_i = \frac{\partial}{\partial x_i}\right)$$

(Thus, E is the "energy operator" of a free particle.) Let G act on H as follows:

$$\rho(g)(\Psi)(x) = \Psi(g^{-1}x) \tag{3.4}$$

for $g \in G$, $x \in R^3$.

(Thus, the action (3.4) corresponds to constructing the product bundle $R^3 \times C$ (C = complex numbers), with the group action also the product). Then, one readily sees that:

$$\rho(g)E = E\rho(g), \tag{3.5}$$

i.e. $\rho(G)$ commutes with E. (This is what is meant by a "symmetry", of course.) Thus, $\rho(G)$ acts on the "energy-levels", i.e. the eigen-functions of E.

Now we will attempt to apply the analysis described in Section 1 to $\rho(G)$.

For $y \in R^3$, set:

$$\Psi_y(x) = e^{ix\cdot y}. \tag{3.6}$$

Thus, $\Psi_y \in H$. Further, for $g_a \in T$,

$$\begin{aligned} \rho(g_a)(\Psi_y)(x) &= \Psi_y(g_a{}^{-1}x) \\ &= \Psi_y(x-a) \\ &= e^{i(x-a)\cdot y} = e^{-ia\cdot y}\Psi_y(x). \end{aligned} \tag{3.7}$$

(3.7) tells us that Ψ_y is an eigenvector of $\rho(T)$, with eigenvalues:

$$\lambda(g_a) = e^{-ia \cdot y}. \qquad (3.8)$$

Hence, we can (and will) use "y" to label points of T^d. This assignment defines an isomorphism of T^d with the additive group[1] of the vector space R^3. Since this additive group is also isomorphic to T, we see that T is "self-dual", i.e. is isomorphic to its dual.

For $y \in R^3$, let H^y denote the space of $\Psi \in H$ such that:

$$\rho(g_a)(\Psi) = e^{-ia \cdot y}(\Psi) \qquad (3.9)$$

for $g_a \in T$.

One proves (exercise) that H^y is one-dimensional, i.e. consists of the constant multiples of Ψ_y

[1]If V is a vector space, it forms an abelian group, where the group-law is the addition operation of the vector space. Further, that we are restricting T^d in this section to be the homomorphisms of T into the complex numbers of absolute value one. This is appropriate because we are basically only studying unitary representations of G.

given by (3.6).

Let us calculate the effect of $\rho(K)$:

$$\rho(k)(\Psi_y)(x) = \Psi_y(k^{-1}x)$$

$$= e^{iy\cdot k^{-1}x} = e^{iky\cdot x}$$

$$= \Psi_{ky}. \tag{3.10}$$

(3.9) and (3.10) determine the effect of $\rho(K)$ and $\rho(T)$ on T^d. Labelling a point of T^d by $y \varepsilon R^3$, we see that:

$$\begin{aligned} g_a(y) &= y \\ k(y) &= ky. \end{aligned} \tag{3.11}$$

Further, the bundle E is just $T^d \times C$, i.e. a cross-section of (E) can be identified with a complex-valued function: $y \to \hat{\Psi}(y)$ [1]. Also, α, the map α: $\Gamma(E) \to H$, can be defined as follows:

---

[1]Since $\hat{\Psi}(y)$ can be identified - as is shown by (3.13) - with the Fourier transform (in the classical sense) of elements $\Psi \varepsilon H$, we choose the notation $\hat{\Psi}$ for elements of $\Gamma(E)$ in order to keep this fact in mind.

$$\alpha(\hat{\Psi}) = \int \hat{\Psi}(y)\Psi_y dy, \tag{3.12}$$

that is, using (3.6),

$$\alpha(\hat{\Psi})(x) = \int \hat{\Psi}(y)e^{ix\cdot y}d^3y. \tag{3.13}$$

Let us determine the action of G on $\Gamma(E)$, in order that the map α defined by (3.12) be intertwining:

$$\rho(g_a)\alpha(\hat{\Psi}) = \int \hat{\Psi}(y)e^{-ia\cdot y}\Psi_y dy.$$

In order that this equals $\alpha(g_a(\hat{\Psi}))$, we must have:

$$g_a(\hat{\Psi})(y) = \hat{\Psi}(y)e^{ia\cdot y} \tag{3.14}$$

for $g_a \in T$.

Also, for $k \in K$,

$$\rho(k)\alpha(\hat{\Psi}) = \int \hat{\Psi}(y)\Psi_{ky}dy$$

$$= \int \hat{\Psi}(k^{-1}y)\Psi_y dy.$$

Thus, $\rho(k)\alpha(\hat{\Psi}) = \alpha(k\hat{\Psi})$ if:

$$k\hat{\Psi}(y) = \hat{\Psi}(k^{-1}y). \tag{3.15}$$

Then, (3.14) and (3.15) determine the action of G on the cross-sections $\Gamma(E)$ of E, hence determine the action of G on E itself. In fact, we can read off this action as follows (exercise):

$$\begin{aligned} &\text{a)} \quad a_a(y, c) = (y, e^{ia\cdot y}c) \\ &\text{b)} \quad k(y, c) = (ky, c) \\ &\qquad \text{for} \quad y \in T^d = R^3, \; c \in C. \end{aligned} \tag{3.16}$$

Then, M, the base-space of the vector bundle E, can be identified with the space of y's, i.e. with R^3. (Again, it should be pointed out that we are restricting attention to "unitary" representations of G, i.e. attempting to decompose H as a sum of "unitary" representations.)

The next step in working out the general theory of Section 1 is to calculate the orbits of G on M. Now, T acts trivially on M, by (3.16a). Thus, it suffices to calculate the orbits of $K = O(3, R)$. These are just the spheres in R^3, of course. If e is a real constant, set:

$$M_e = \{y \in R^3: \ y \cdot y = 2e^2\}.$$

Define α_e: $F(M_e) \to H$ as follows:

$$\alpha_e(\hat{\Psi})(x) = \int_{M_e} \hat{\Psi}(y)e^{ix \cdot y} d_e y. \qquad (3.17)$$

Here, $y \to \hat{\Psi}(y)$ is a function on M_e, $d_e y$ is a volume element differential form on the manifold M_e that is invariant under the action of K. (This is just the usual rotation - symmetric volume element on the sphere, of course.) Then, one readily computes that:

$$E^2(\alpha_e(\hat{\Psi})) = e^2\alpha(\hat{\Psi}), \qquad (3.18)$$

i.e. $\alpha_e(F(M_e))$ has in the space of eigenvectors of $E = \frac{1}{2} \Delta^2$, the "Hamiltonian" or "energy operator" of a free-particle moving in R^3.

Conversely, one can show - by using the inverse Fourier transform - that $\alpha_e(F(M_e))$ essentially[1] fills up this energy level. This (using a

[1]We are being vague at this point because we do not want to go into the technicalities of Hilbert and generalized-function theory to describe the precise result.

suitably generalized version of Theorem 1.1) indicates that $\rho(G)$ acts irreducibly on the energy levels.

Finally, we can indicate how H can be written as a "direct integral" of the subspace $\alpha(F(M_e))$. Notice that there is a function h(e) of one-variable such that:

$$\int \hat{\Psi}(y)d^3y = \int_0^\infty \left(\int_{M_e} \hat{\Psi}(y)d_e y\right)h(e)de \qquad (3.19)$$

for $\hat{\Psi} \varepsilon \Gamma(E)$.

This indicates that we are writing a $\hat{\Psi} \varepsilon \Gamma(E)$ a "direct integral":

$$\hat{\Psi} = \int_0^\infty \hat{\Psi}_e h(e)de, \qquad (3.20)$$

where $\hat{\Psi}_e(y) = \hat{\Psi}(y)\delta(y\cdot y - 2c^2)$. Applying α to both sides of (3.20) gives a corresponding "direct integral" decomposition of $\alpha(\Gamma(E))$ in terms of the $\alpha_e(F(M_e))$. (Thus, a function $\hat{\Psi}_e \varepsilon F(M_e)$ is identified with the "generalized function"

$$y \to \hat{\Psi}_e(y)\delta(y\cdot y - 2c^2)$$

in terms of $\Gamma(E)$.

Now, we turn to another example - the "crystallographic groups", which are subgroups of $G = E(3)$.

4. THE CRYSTALLOGRAPHIC GROUPS

We must first describe the geometric terminology needed to determine the "crystallographic groups" and "crystals".

Now, R^3 is a real, three-dimensional vector space. We are denoting points of R^3 by "x". Let (x_1, x_2, x_3) denote an arbitrary basis of R^3.

DEFINITION. A *lattice* (or *crystal*) in R^3 (associated with a basis (x_1, x_2, x_3)) is the set of points $x \in R^3$ that can be written in the form:

$$x = m_1x_1 + m_2x_2 + m_3x_3 \tag{4.1}$$

where (m_1, m_2, m_3) are integers

Given such a lattice, the subgroup of $g \in G = E(3)$ that preserve the lattice, i.e. that map x_1, x_2, x_3, into points of form (4.1), is called the *group* of the lattice. A subgroup G_0 of G is called a *crystallographic group* it if preserves *some* lattice.

The crystallographic groups have been classified for a long time; the remarkable fact is that there are very few of them, hence the crystals that occur in nature may be classified into families according to the types of group they admit. (We will use Tinkham [1] as a general reference in this area, particularly for its references to further mathematical and physical literature.)

Given such a crystal group G_0, the intersection subgroup $G_0 \cap K = K_0$ is called the *point group* of the lattice, and the subgroup $G_0 \cap T = T_0$ is called the *translation subgroup* of the lattice. Notice that K_0 has the following properties:

a) K_0 is a finite subgroup of $K = O(3, R)$ (4.2)

b) K_0 has at least one representation by 3×3 matrices, all of whose entries are integers.

In fact, one can classify all subgroups of

O(3, R) satisfying a) (see J. Wolf [1]). There turn out to be five equivalence classes of such groups ("equivalence" means that they are conjugate under automorphisms of O(3, R)); each class can be realized as a symmetry group of a "perfect solid". All but one of these groups (the largest, the "octahedral group", does not leave a lattice invariant) also satisfy (4.2b), i.e. appear as point groups associated with a crystal group.

Now, we have automatically:

$$K_C \cap T_0 = (\text{identity})$$

However, it does not necessarily follow that

$$G_0 = K_0 \cdot T_0, \tag{4.3}$$

i.e., that G_0 is the semidirect product of K_0 and T_0. However, most of the crystals that occur in nature do satisfy (4.3), and we will work with this condition.

We must now describe Poincaré's geometric construction of a "fundamental domain" associated with a discrete subgroup of G.

DEFINITION. A subgroup S of G(= E(3)) is *discrete* if:

a) S has at most a countable number of elements.

b) The elements of S have no accummulation points within G, i.e. there is no sequence $g_0, g_1, \ldots$ of *distinct* elements of G which converge to an element of G.

Consider a point $x_0 \in R^3$, and the orbit: $N = Sx_0$: of such a discrete subgroup S. Let D be the subset of R^3 defined as follows:

$$x \in D \text{ if and only if } |x - x_0| \leq |x - gx_0| \quad (4.4)$$

for all points gx_0:

We must now describe the properties of D relative to the action of S. First we have:

THEOREM 4.1. Given a point $x \in R^3$ there is a $g \in S$ such that

$$gx \in D$$

(In other words, $SD = R^3$)

Proof. First, one shows that N, as a subset of R^3, is a closed set of points (exercise). Then, given $x \in R^3$, there is a point $g_0 x_0 \in N$, with $g_0 \in S$, which is "closest" to x, i.e.

$$|x - g_0 x_0| \leq |x - y| \tag{4.6}$$

for all $y \in N$

Now, since S is a subgroup of $G = E(3)$,

$$|gx - gy| = |x - y| \tag{4.7}$$

for $x, y \in R^3$, $g \in S$

Applying this to (4.6) proves:

$$|g_0^{-1} x - x_0| \leq |g_0 x - g_0^{-1} y|$$

for all $y \in N$

Since $g_0 N \subset N$ (since N is an orbit) this shows that $g_0^{-1} x$ satisfies (4.4), hence

$g_0^{-1} x \in D$, i.e.

$$x = g_0(g_0^{-1}x) \in g_0 D$$

This proves (4.5).

The second basic geometric property of D can be described as follows:

<u>THEOREM 4.2</u>. Let D_0 be the "interior" of D, i.e. the set of points $x \in R^3$ such that

$$|x - x_0| < |x - y| \qquad (4.8)$$

for all points $y \in N$ which are not equal to x_0.

Then, if $g \in S$ is not the identity, and if $x \in R^3$, then $gx \notin D$.

<u>Proof</u>. Suppose otherwise, i.e. $gx \in D$. Then, $|x - x_0| < |x - y|$

$$|gx - x_0| \leq |gx - y| \qquad (4.9)$$

for all $y \in N$

Using (4.7), $|x - x_0| = |gx - gx_0|$,

$$|gx - gx_0| < |gx - gy| \qquad (4.10)$$

for all $y \in N$

Since x_0 and $gx_0 \in N$, note that (4.9) and (4.10) are contradictory. □

The general properties of D covered by the conclusions of Theorems 4.1 and 4.2 lead to calling D a *fundamental domain* for the action of S. Restating then, we see that each orbit of S touches D at least once, and if an orbit touches D_0, it has a *unique* representative in D. The orbits that touch the boundary $D - D_0$ may have several representative points in D. Further, these results are not dependent on the Euclidean nature of R^3, but hold if S is an arbitrary, discrete group of isometries of a Riemannian manifold. This more general case is important for the study of the discrete groups occurring in the theory of automorphic functions.

Let us return to the study of a crystal group G_0, satisfying (4.3). One may verify (exercise) that it is a discrete subgroup of $G = E(3)$. Let us study its "irreducible" representations, constructed by the methods of Section 1. Now, T_0 is generated by three elements g_1, g_2, g_3 with

$$g_i(x_j) = x_j + \delta_{ij} \tag{4.11}$$

for $1 \leq i, j \leq 3$.

Thus, an arbitrary $g \in T_0$ may be written as:

$$g_{m_1, m_2, m_3} = g_1^{m_1} g_2^{m_2} g_3^{m_3} \tag{4.12}$$

with integers m_1, m_2, m_3.

Now, let T_0^d[1] be the "dual" group of T_0 defined as the homomorphisms λ: $T_0 \to C$, such that:

$$|\lambda(g)| = 1 \quad \text{for all} \quad g \in T_0. \qquad (4.12)$$

This can be conveniently parameterized in terms of the "dual lattice".

<u>DEFINITION</u>. Given a basis (x_i) of R^3, the vectors

---

[1] Again - since we are basically only interested in unitary representations - we are considering T^d to be the homomorphisms into the multiplicative group of complex members of absolute value one.

(y_i) such that:

$$x_i \cdot y_j = 2\pi\delta_{ij} \tag{4.13}$$

are called the dual basis[1] of R^3.

The lattice generated by (y_1, y_2, y_3) is called the *dual lattice* of the lattice generated by (x_1, x_2, x_3).

<u>THEOREM 4.3</u>. Let T_0' be the translation subgroup associated with the dual lattice. Then, the dual group T_0^d is isomorphic to the quotient group T/T_0', where T is the group of all translations of R^3.

<u>Proof</u>. Given $g_a \in T$, with $a \in R^3$, associated with g_a the element λ_a: $T \to C$, as follows:

$$\lambda_a(g_{m_1,m_2,m_3}) = e^{i(m_1(a\cdot x_1)+m_2(a\cdot x_2)+m_3(a\cdot x_3))} \tag{4.14}$$

It is obvious from (4.14) that, when a belongs to

[1]Exercise: Show that (y_1, y_2, y_3) indeed form a basis of R^3.

the dual lattice, λ_a = identity. Then $a \rightarrow \lambda_a$ passes to the quotient to define a homomorphism: $T/T_0{}' \rightarrow T_0{}^d$. It is left as an exercise to show that this homomorphism is *the* isomorphism between $T/T_0{}'$ and $T_0{}^d$, needed to prove the theorem.

This result-together with the ideas of Section 1 - enables us to describe the irreducible representations of G_0. Recall that the first step is to describe the orbits of K_0 acting on $T_0{}^d$. Now, with the isomorphism (4.14), it is readily seen that the action of K_0 on $T_0{}^d$ is obtained from the action of K on T by passing to the quotient. Since K leaves invariant the lattice generated by (x_1, x_2, x_3), it also leaves invariant the dual lattice (exercise), i.e. given $k \in K$, the map $g_a \rightarrow hg_ak^{-1} = g_{ka}$ map to $T_0{}'$ into itself, thus defining an automorphism on the quotient group $T/T_0{}'$. Our job is to find the orbits of K acting on $T/T_0{}'$.

<u>THEOREM 4.4</u>. Let $G_0{}'$ be the semidirect product group: $K\cdot T_0{}'$. Then the orbits of K acting on $T/T_0{}' = T_0{}^d$ are in one-one correspondence with the orbits of $G_0{}'$ acting on R^3.

Proof. Identify T with R^3, and the map $a \to g_a$, where $a \in R^3$. Then, $T_0{}'$ consists of the $g_{a_0}{}'$, where $a_0{}'$ belongs to the dual lattice, generated by (y_1, y_2, y_3). An orbit of $G_0{}'$ acting on a consists of all vectors of the form:

$$k(a) + a_0{}', \tag{4.15}$$

where $G \in K$, $a_0{}'$ belongs to the dual lattice.

On the other hand, the inverse image under the projection map: $R^3 = T \to T/T_0{}'$ of an orbit of K consists precisely of a set of elements of form (4.15). This proves the theorem.

Next, notice that Poincaré's fundamental domain construction - applied to the group $G_0{}'$ acting on R^3 - gives a convenient way of parameterizing the orbits of $G_0{}'$, hence also - by Theorem 4.4 - of the orbits of K acting on $T_0{}'^d$.

Finally, to complete the identification of the special features of this problem with the general theory treated in Section 1, it is necessary to identify the isotropy subgroup of K acting on $T_0{}^d$. For this purpose we can use Theorem 4.3 again, identifying $T_0{}^d$ with $T/T_0{}'$. Suppose

$a \varepsilon R^3$, g_a T. Then, clearly k acting on $T/T_0{}'$ leaves the image of g_a in $T/T_0{}'$ fixed if and only if:

$$k(a) - a \text{ belongs to the dual lattice} \quad (4.16)$$

Then, the isotropy subgroup of K at a point in $T/T_0{}' = T_0{}^d$ represented by g_a is the set of $k \varepsilon K$ satisfying (4.16), which we may denote by K^a.

We can then recapitulate in more purely group-theoretic language how induced representation of G_o may be constructed. Pick a point $a \varepsilon R^3$. Let K^a be the subgroup of the point-group K satisfying (4.16). Let $\sigma(K^a)$ be representation of K^a by operators on a vector space V. Extend σ to be a representation of the subgroup $K^a \cdot T_0$, in the following way:

$$\sigma(kg_b) = \sigma(k)e^{ia\cdot b} \quad (4.17)$$

for $g_b \varepsilon T_0$

with b an element of the lattice generated by (x_1, x_2, x_3), and with $k \varepsilon K^a$.

Then, construct a representation ρ of G by "inducing" σ from the subgroup $K^a \cdot T_0$ to G, i.e. use σ to construct homogeneous vector bundle: $E \to G/K^a \cdot T_0$, and let G act on the cross-sections $\Gamma(E)$.

As a check on the consistency of what we have done, let us verify independently of what we have already done that (4.17) defines σ as a genuine representation of $K \cdot T_0$. Thus, for k, $k' \in K^a$, b,b' in the lattice

$$\begin{aligned} &\sigma(kg_b k' g_{b'}) \\ &= \sigma(kk'(k'^{-1} g_b k') g_{b'}) \\ &= \sigma(kk' g_{k'^{-1}(b)} g_{b'}) \qquad (4.18) \\ &= \sigma(kk' g_{k'^{-1}(b)+b'}) \\ &= \text{, using (4.17), } \sigma(kk') e^{ia\cdot(k'^{-1}(b)+b')} \end{aligned}$$

On the other hand, if σ defined by (4.17) is to be a genuine representation, one must have

$$\begin{aligned} &\sigma(kg_b k' g_{b'}) = \sigma(kg_b)(k' g_{b'}) \\ &= \text{, using (4.17)} \\ &\sigma(k)\sigma(k') e^{ia\cdot b}\, e^{ia\cdot b'} \qquad (4.19) \end{aligned}$$

Let us work on the right hand side of (4.18), to show that it is equal to (4.19). Now, using (4.16)

$$a\cdot k'^{-1}(b)$$
$$= k'(a)\cdot b \qquad (4.20)$$
$$= a\cdot b + (\text{a vector in the dual lattice})\cdot b$$

Now, b is a vector in the lattice generated by (x_1, x_2, x_3). Hence, the second term on the right hand side of (4.20) is - by the definition (4.13) of the "dual lattice" - an integer multiple of 2π, hence is equal to 1 when it is multiplied by $\sqrt{-1} = i$ and exponentiated. This shows independently that (4.18) equals (4.19), hence that σ defines a genuine representation of $K^a\cdot T_0$, which in turn can be "induced" to define a representation ρ of G on the cross-sections $\Gamma(E)$ of the homogeneous vector bundle on $G/K^a\cdot T_0 = K/K^a$.

We can also describe briefly how these ideas are used in solid-state physics. Let H be the vector space of complex-valued, C^∞ functions $x \to \Psi(x)$ which are "periodic" on the lattice, i.e.

$$\Psi(x + a) = \Psi(x) \qquad (4.21)$$

for $x \in R^3$, a in the lattice.

(Thus, H are the C^∞ functions on the manifold T/T_0. Topologists call this manifold a *torus*). Let $V(x)$ be a real-valued function satisfying (4.21). Let E be the linear operator: $H \to H$ defined as follows:

$$E(\Psi)(x) = 1/2(\Delta\Psi)(x) + V(x)\Psi(x), \qquad (4.22)$$

where $\Delta = \frac{\partial}{\partial x_i}\frac{\partial}{\partial x_i}$ is the Laplace operator, $V(x)$ is a "periodic potential". Thus, (4.22) describes "particles" moving on the lattice, subject to the potential V.

Define the representation $\rho(G_0)$ of G_0 by operators on H as follows:

$$\rho(g)(\Psi)(x) = \Psi(g^{-1}x)$$

for $x \in R^3$, $\Psi \in H$, $g \in G_0$

Now, as we have seen in Section 3, $\rho(g)$ commutes with Δ. If $g \in T$, $\rho(g)$ also commutes with E. In solid-state physics, one typically chooses V so that either: E commutes with $\rho(K)$ also, i.e. V has the symmetry of the "point group", (obviously the simplest case), or, V belongs to a set of such possible potentials which transform among themselves under $\rho(K)$ (the case where this set forms an irreducible representation of K would be the next simplest case).

The next step is obviously to decompose $\rho(G)$ acting on H into irreducible representation. This follows the pattern established by G = E(3) in Section 2. Then, one uses various perturbation-theoretic methods (the "standard" perturbation methods of quantum mechanics) to pull out various bits and pieces of information (mostly qualitative in nature) concerning the "relation between the energy levels" i.e. the eigenvalues of E, and the irreducible representations of G. We will not go into this material at this stage. We may note, however, that this group-theoretic approach leads to remarkable mathematical analogies between solid state physics and relativistic elementary particle physics - the basic group $G_0 \subset E(3)$ is replaced by the inhomogeneous Lorentz group, i.e. the "Poincaré group". Presumably, this analogy was basic in Wigner's mind in his emphasis (and that of the physicists he influenced) on the role that the

Poincaré group plays in elementary particle physics, since he and his students had earlier done in the 1930's some of the basic work sketched above concerning the representations of the crystallographic groups.

Let us now turn to the study of the Poincaré group.

5. REPRESENTATIONS OF THE POINCARÉ GROUP

We must now change our notations, since our basic "space" changes from R^3 to R^4. Let x now denote a point of R^4. Choose indices as follows:

$$0 \leq \mu, \nu \leq 3, \ 1 \leq i, j \leq 3$$

$$x = (x_\mu)$$

Let $(g_{\mu\nu})$ be the Lorentz metric tensor

$$g_{00} = 1; \quad g_{ij} = -1$$

$$g_{0i} = g_{i0} = 0$$

For x, y ε R^4, set:

$$x \cdot y = g_{\mu\nu} x_\mu y_\nu .$$

i.e. $x \cdot y$ is the Lorentz inner product on R^4. $(x, y) \to x \cdot y$ is then a symmetric, bilinear form and the real vector space R^4.

DEFINITION. A *translation* is a map

$g: R^4 \to R^4$ of the form:

$g(x) = x + a$, for $x \in R^4$,

where a is a fixed vector in R^4. A typical such translation is denoted by g_a. The set of such transformations forms a group, called the *translation group*, denoted by "T". The assignment $a \to g_a$ defines an isomorphism between T and the "additive group" of R^4, i.e. the set of 4-vectors, with the group law defined by "addition" of vectors.

A *Lorentz transformation*, typically denoted by "ℓ", is a linear transformation: $R^4 \to R^4$ such that:

$$(\ell x) \cdot (\ell y) = x \cdot y$$

for $x, y \in R^4$.

For $\ell \in L$, $g_a \in T$, $a \in R^4$ note that:

$$\ell g_a \ell^{-1} = g_{\ell(a)} \tag{5.1}$$

(5.1) shows that the map $g_a \to \ell g_a \ell^{-1}$ defines an isomorphism of T; as ℓ varies, this defines a homomorphism of L into the group of isomorphism of T. The *semidirect product* $G = L \cdot T$ defined by this homomorphism is called the *Poincaré group*.

Of course, alternately, G can be defined as the set of transformations $g\colon\ R^4 \to R^4$ which can be written as products:

$$g = \ell g_a, \tag{5.2}$$

for *some* $\ell \in L$, $g_a \in T$.

Then, relation (5.1) shows that the collection of such transformations defines a group, which we may identify with the Poincaré group defined "abstractly" above. (*Exercise:* Carry out the details of this identification.)

The next step in the construction of representations of G by the method outlined in Section 1 is to make explicit a description of the dual group T^d. This may be done as follows:

Introduce a "new" four vector $p = (p_\mu) \varepsilon R^4$. (Ultimately, "p" will be identified with the "energy-momentum four-vectors" of particles; this explains our choice of notation, since "p" is standard notation in the physics literature for the "energy-momentum four vector"). Given such a p, define:

$$\lambda_p\colon \ T \rightarrow C$$

as follows:

$$\lambda_p(g_a) = e^{ip \cdot a} \tag{5.3}$$

Notice that $\lambda_p \varepsilon T^d$, i.e. λ_p is a homomorphism of T into the multiplicative group of complex numbers (of absolute value one, since we are only interested - in this chapter - in unitary representations.)

With this identification of T^d with R^4 (again, the group T is "self-dual"), we may proceed to calculate the orbits of L on T^d - the essential first step in describing and constructing the irreducible representations.

Now given $\ell \in L$,

$$(\ell\lambda_p)(g_a) = \lambda_p(\ell^{-1}g_a\ell)$$

$$= \lambda_p(g_{\ell^{-1}a})$$

$$= e^{i\ell^{-1}a \cdot p} = e^{ia \cdot \ell p}$$

$$= \lambda_{\ell(p)}(g_a).$$

Thus,

$$(\ell\lambda_p) = \lambda_{\ell p} \tag{5.4}$$

With this identification of T with R^4, we see that the action of L goes over into the action of L on R^4 by which L is defined, $p \to \ell p$.

To find the orbits of L on R^4, it is most convenient to use the following result, the "Witt theorem." (See Wolf [1] for the proof).

<u>THEOREM 5.1</u>. Let V be a vector space over the real numbers, and let $(v, v') \to \beta(v, v')$ be a

symmetric, bilinear form on V. Suppose $(v_1, \ldots, v_n)$ and $(v_1', \ldots, v_n')$ are two sets of linearly independent elements of V such that:

$$\beta(v_i, v_j) = \beta(v_i', v_j') \tag{5.5}$$

for $1 \leq i, j \leq n$

Then, there is a linear transformation $A: V \to V$, such that:

$$\text{a)} \quad A(v_i) = v_i'$$
$$\text{b)} \quad \beta(Av, Av') = \beta(v, v') \text{ for } v, v' \in V \tag{5.6}$$

We can apply this to the case: $V = R^4$; β is the form defined by the Lorentz inner product; and $n = 1$. Given $p \in R^4$, let me be its "mass", i.e.

$$m^2 = p \cdot p$$

L is the set of linear transformations of R^4 satisfying (5.6b). We see from (5.6a) that there is an $\ell \in L$ mapping p onto p' (if $p, p' \neq 0$) if and only if:

$$p^2 = p'^2$$

Then, the orbits of L are the "mass-hyperboloids" of R^4, i.e. the manifolds M(m) defined as follows:

$$M(m) = \{p \in R^4 : \ p^2 = m^2\} \tag{5.7}$$

The isotropy subgroup of L at points of M(m) can be readily calculated.

Let (p_μ) be a basis of R^4 such that

$$g_{\mu\nu} = p_\mu \cdot p_\nu \tag{5.8}$$

If A: $R^4 \to R^4$ is a linear transformation, define:

$$A_{\mu\nu} = p_\mu \cdot A p_\nu \tag{5.9}$$

Then, $(A_{\mu\nu})$ is the 4 × 4 real matrix corresponding to A, defined by the basis (p_μ).

<u>Case 1</u>. $m^2 > 0$

Then, $mp_0 \in M(m)$. Let K be the isotropy subgroup of L at mp_0, i.e. the set of $\ell \in L$ such that:

$$\ell(mp_0) = Mp_0, \quad \text{or} \quad \ell(p_0) = p_0 \tag{5.10}$$

Then,

$$\ell_{\mu 0} = p_\mu \cdot \ell p_0 = p_\mu \cdot p_0 = g_{\mu 0} \tag{5.11}$$

It also follows from (5.10) that:

$$\ell^{-1} p_0 = p_0$$

Hence,

$$\ell_{0\mu} = p_0 \ell p_\mu = \ell^{-1} p_0 p_\mu = g_{\mu 0} \tag{5.12}$$

Then, (5.11) and (5.12) are the conditions that $\ell_{0\mu} = p_0 \cdot \ell p_\mu = \ell^{-1} p_0 \cdot p_\mu = g_{\mu 0}$ belong to K. Notice that $(\ell_{\mu\nu})$ is of the following matrix form:

$$(\ell_{\mu\nu}) = \begin{pmatrix} 1, & 0, & 0, & 0 \\ 0, & & & \\ 0, & & \ell_{ij} & \\ 0, & & & \end{pmatrix}$$

Thus, the mapping $\ell \to (\ell_{ij})$ defines a homomorphism of K onto a group - in fact, the group $O(3, R)$ of 3×3 orthogonal matrices.

Another more "intrinsic" way of seeing this can be described as follows: Let V be the orthogonal complement of p_0 in R^4, i.e. the set of vectors $p \in R^4$ such that:

$$p \cdot p_0 = 0$$

Given $\ell \in K$, since it leaves p_0 invariant, it maps $V \to V$. It leaves invariant the form resulting from restricting the Lorentz inner product to V. However, this form is negative definite. By the Witt theorem, any linear transformation mapping V into V, preserving the Lorentz product, can be extended to a Lorentz transformation leaving p_0 invariant, i.e. to an element of K. This shows that there is a isomorphism: $K \to$ (linear transformations on V) leaving invariant the negative-definite form).

This algebraic argument will be useful in more complicated situations.

<u>Case 2</u>. $m^2 < 0$.

Then if $a^2 = -m^2$, $(ap_2)^2 = m^2$, i.e. $ap \in M(m)$.

Then we can define K' as the isotropy subgroup of L at p_1.

To calculate K', let V be the orthogonal complement of p_1 in R^4.

It has a basis consisting of (p_0, p_2, p_3). Then, the Lorentz inner product restricted to V is a symmetric bilinear form, with one plus sign, and two minus signs. The group of its linear automorphism is denoted by: O(1, 2):

Again, $\ell \in$ K' maps V into itself, and the Witt theorem implies that the mapping $\ell \to$ (group of linear transformations on V) defines an isomorphism between K' and O(1, 2).

Case 3. $m^2 = 0$.

Notice now that: $(p_0 + p_1)^2 = p_0{}^2 + p_1{}^2 = 0$, i.e. $p_0 + p_1 \in M(0)$. Let K" be the isotropy subgroup of L at $(p_0 + p_1)$. The structure of K" is less obvious than was that of K or K'. We will now investigate it.

The result is that K" is isomorphic to E(2), the group of rigid motions of Euclidean 2-space. In fact, we will only prove here a weaker version of this - that its Lie algebra is isomorphic to the Lie algebra of E(2) - leaving it as an exercise to prove the group-statement.

It will be useful to make explicit a basis for the Lie algebra of L, denoted by $\underset{\sim}{L}$, in terms of the basis (p_μ) for R^4. Now, by general principles, L denotes the Lie algebra (under commutator) of linear transformations A: $R^4 \to R^4$ such that:

$$Ap \cdot p' = - p \cdot Ap' \tag{5.13}$$

for $p, p' \in R^4$

Define linear transformations (J_i, K_i) as follows:

$$\begin{aligned} J_i(p_0) &= 0 \\ J_i(p_j) &= \varepsilon_{ijk} p_k \end{aligned} \tag{5.14}$$

$$\begin{aligned} K_i(p_0) &= p_i \\ K_i(p_j) &= \delta_{ij} p_0 \end{aligned} \tag{5-15}$$

Then, one shows readily (exercises) that:

a) J_i, K_i are in $\underset{\sim}{L}$, i.e. satisfy (5.13)

b) (J_i, K_i) form a basis for L.

c) This basis satisfies the following structure relations:

$$
\begin{aligned}
[J_i, J_j] &= \varepsilon_{ijk} J_k \\
[J_i, K_i] &= \varepsilon_{ijk} K_k \qquad (5.16) \\
[K_i, K_j] &= \varepsilon_{ijk} J_k
\end{aligned}
$$

Then we recognize that:

a) The (J_i) form a basis for $\underset{\sim}{K}$.

b) J_1, K_2, K_3 forms a basis for $\underset{\sim}{K}'$.

We shall now show that:

$$
(J_1, J_2 + K_3, J_3 - K_2) \qquad (5.17)
$$

is a basis for $\underset{\sim}{K}''$

<u>Proof</u>. It can be proved from dimensional considerations (exercise), that K", hence $\underset{\sim}{K}''$, is three dimensional. It should also be clear that the elements listed in (5.17) are linearly independent. Then - to prove (5.17) - it suffices to show that these three elements actually belong to

K", i.e. that, when applied to $p_0 + p_1$, they give zero.

Using (5.14) and (5.15),

$$J_1(p_0 + p_1) = 0$$

$$(J_2 + K_3)(p_0 + p_1) = J_2p_0 + J_2p_1$$

$$+ K_3p_0 + K_3p_1 = - p_3 + p_3 = 0$$

$$(J_3 - K_2)(p_0 + p_1) = J_3p_0 + J_3p_1$$

$$- K_2p_0 - K_2p_1 = p_2 - p_2 = 0.$$

Now, let us use (5.16) to show that the Lie algebra generated by $(J_1, J_2 + K_3, J_3 - K_2) = K''$ is isomorphic to the Lie algebra of E(2). Now, E(2) is the semi-direct product of O(2, R) and the abelian two-dimensional translation group, isomorphic to R^2. Then, its Lie algebra is a semi-direct sum of a two dimensional abelian ideal, and a one dimensional subalgebra, (the Lie algebra of the O(2, R) subgroup).

(*Exercise*: Show that E(2) itself is a *solvable* group.) We will show that K" has a similar structure; using (5.16):

$$[J_1, J_2 + K_3] = J_3 - K_3$$

$$[J_1, J_3 - K_2] = - J_2 - K_3 = - (J_2 + K_3)$$

$$[J_2 + K_3, J_3 - K_2] = [J_2, J_3] \tag{5.18}$$

$$- [K_3, K_2] = J_1 - J_1 = 0$$

(5.18) now shows that K" has the desired structure.

Now, we have finished the description of the orbits of L on T^d, and the calculation of the isotropy subgroups at these orbits. In principle, this enables us to apply the general theory of Section 1 and, in terms of the theory of vector bundles and induced representations, write down the irreducible representations of G. Indeed, these results immediately show us that these representations are labelled by two parameters (m, s), where "m" is the "mass", "s" is the spin. In fact, m can be identified with the number m labelling the orbit M(m), while "s" is the parameter labelling the irreducible representation σ of the isotropy subgroup of the L on the orbit M(m) with which one constructs the vector bundle E over M(m) whose cross-sections $\Gamma(E)$ are identified with the vector space on which G sets.

However, rather than proceed directly in this purely group-theoretic direction at this moment, we prefer to pause in order to develop the more traditional material concerning "Lorentz invariant wave equations."

CHAPTER VII

POINCARÉ-INVARIANT DIFFERENTIAL OPERATORS AND BUNDLES

In Chapter VI, we have investigated group-theoretic methods of description of representations of the Poincaré group. However, the representations that appear naturally in quantum field theory are usually associated with Poincaré invariant partial differential operators on Minkowski space. The study of these equations is also essential for a description of free quantum fields of various spins. In this chapter, we will systematically study such operators, using - as a general reference - Gel'fand, Minlos and Shapiro [1] and Naimark [1].

1. POINCARÉ-HOMOGENEOUS VECTOR BUNDLES ON R^4

Now, R^4 denotes physical "space-time". (It may also of course be used to denote the space of "energy-momentum vectors", which is "dual" to "space-time".) Points of R^4 will be denoted by "x", with $x = (x_\mu)$, $0 \leq \mu, \nu \leq 3$, denoting its components.

$$x \cdot y = g_{\mu\nu} x_\mu y_\nu \tag{1.1}$$

denotes the Lorentz inner product. ∂_μ denotes the differential operator $\frac{\partial}{\partial x_\mu}$.

Let G be the Poincaré group, considered as a transformation group on R^4. Recall the notations of Chapter VI: T denotes the translation; $g_a \in T$, for $a \in R^4$, with:

$$g_a(x) = x + a.$$

L denotes the linear, Lorentz-transformations: $R^4 \to R^4$. G is the semidirect product: L·T.

Let (E, R^4, π) be a vector bundle over R^4, which is acted on linearly by G, with the action on the base R^4 just the "natural" linear action.

Let 0 be the element "zero" of R^4, and let $V = E(0)$ be the fiber of E over 0. We will show that E can be exhibited as isomorphic to the product $R^4 \times V$.

Define a map α: $E \to R^4 \times V$, as follows: Given a point $x \in R^4$, $v \in E(x)$, set:

$$\alpha(v) = (x, g_{-x}(v)),$$

where $v \to g_x v$ denotes the linear map: $E \to E$ associated with the translation g_x.

It should be clear that α is the desired vector bundle isomorphism. Let us compute the action of G on $E \times V$ required to make α an intertwining map:

For $g_a \in T$, $v \in E(x)$,

$$\alpha(g_a(v)) = (x+a, g_{-x-a}g_a(v)) = (x+a, g_{-x}(v).$$

Thus, if we define g_a: $R^4 \times V \to R^4 \times V$ as follows, α will be intertwining for the action of T:

$$g_a(x, v) = (x + a, v). \qquad (1.2)$$

(1.2) can be described as saying that T acts

on the base R^4 as it acts geometrically, but acts "trivially" in a "non-twisted" way on the fiber V.

However, the action of L is not this simple, in general. Now L is the isotropy subgroup of G acting on R^4, at the point $x = 0$. Let $\sigma(L)$ be the representation of L by operators acting on the fiber $V = E(0)$ (the "linear isotropy representation" associated with the action of G on the vector bundle).

Define L acting on $R^4 \times V$, as follows:

$$\ell(x, v) = (\ell x, \sigma(\ell)(v)) \tag{1.3}$$

for $x \in R^4$, $v \in V$.

We shall now show that the action L on $R^4 \times V$ is the appropriate one in order that α intertwine the action of L:

For $x \in R^4$, $v \in E(x)$,

$$\alpha(\ell v) = (\ell x, g_{-\ell x}(\ell v))$$

$$= (\ell x, \ell(\ell^{-1} g_{-\ell x}(\ell)(v))$$

$$= (\ell x, \ell g_{-x}(v))$$

$$= (\ell x, \sigma(\ell)(g_{-x}(v)). \tag{1.4}$$

On the other hand, using (1.3),

$$\ell(\alpha(v)) = \ell(x, g_{-x}(v))$$

$$= (\ell x, \ell g_{-x}(v)). \qquad (1.5)$$

(1.4) and (1.5) are equal, which shows that α intertwines the action of L. Since we have shown already that it intertwines T, we see that we can replace E, with the given action of G, with $R^4 \times V$, with the action of T and L on $R^4 \times V$ given by (1.2) and (1.4).

Having identified E with a product, we can identify $\Gamma(E)$ as the space of all maps Ψ: $R^4 \to V$. The representation ρ of G on the cross-sections then translates - using (1.2) and (1.3) - over to the following action on this space of all maps:

$$\text{a)} \quad \rho(g_a)(\Psi)(x) = \Psi(x-a)$$
$$\text{for} \quad \Psi \in \Gamma(E),\ x \in R^4,\ g_a \in T$$
$$(1.6)$$
$$\text{b)} \quad \rho(\ell)(\Psi)(x) = \sigma(\ell)\Psi(\ell^{-1}x)$$
$$\text{for} \quad \ell \in L,\ \Psi \in \Gamma(E),\ x \in R^4.$$

These are the rules used in the physicist's

approach to the subject - for example, in the treatment of the Dirac equation in quantum field theory textbooks. (For example, Schweber [1], Bjorken and Drell [1], Bogoliubov and Shirkov [1] and Jost [1].) Of course, the simple geometric nature of Minkowski space - all Poincaré homogeneous vector bundles are products - enables physicists to use the rules (1.6) in a perfectly consistent way, while remaining in ignorance about the general underlying theory of vector bundles.

2. POINCARÉ-INVARIANT FIRST ORDER DIFFERENTIAL OPERATORS

Suppose that E is a Poincaré-homogeneous vector bundle on Minkowski space, with $\Gamma(E)$ identified with the space of all maps Ψ: $R^4 \to V$, as explained in Section 1, and with the action $\rho(G)$ of the Poincaré group on $\Gamma(E)$ defined by (1.6). Our aim is to determine all first-order, linear, differential operators D: $\Gamma(E) \to \Gamma(E)$ which commute with $\rho(G)$.

As a first-order, linear differential operator, D must be of the following form:

$$(D\Psi)(x) = \gamma^{\mu}(x)\partial_{\mu}\Psi(x) + \gamma(x)\Psi(x), \qquad (2.1)$$

where, for $x \in R^4$, $\gamma^{\mu}(x)$, $\gamma(x)$ are linear transformations: $V \to V$.

To investigate the conditions that:

$$\rho(g)D(\Psi) = D\rho(g)(\Psi) \qquad (2.2)$$

for all $g \in G$

we can first consider the case where $g \in T$. We see from (1.6a) that (2.2) will be true for all $a \in R^4$ if and only if:

$$\gamma^{\mu}(x), \gamma(x) \text{ are independent of } x, \qquad (2.3)$$

i.e. a T-invariant, first order linear differential operator is determined by five linear transformations γ^{μ}, γ: $V \to V$.

Given (2.3), let us investigate (2.2), for $g = \ell \in L$.

$$\rho(\ell)(D(\Psi))(x) = \sigma(\ell)(D(\Psi)(\ell^{-1}x))$$

$$= \sigma(\ell)(\gamma^{\mu}\partial_{\mu}(\Psi)(\ell^{-1}x) + \gamma\Psi(\ell^{-1}x))$$

$$= [\sigma(\ell)\gamma^{\mu}\sigma(\ell^{-1})](\rho(\ell)(\partial_{\mu}\Psi)(x))$$

$$+ [\sigma(\ell)\gamma\sigma(\ell^{-1})]\rho(\ell)(\Psi)(x).$$

Thus,

$$\rho(\ell)D(\Psi) = [\sigma(\ell)\gamma^{\mu}\sigma(\ell^{-1})]\rho(\ell)\partial_{\mu}\Psi + [\sigma(\ell)\gamma\sigma(\ell^{-1})]\rho(\ell)\Psi. \tag{2.4}$$

On the other hand,

$$D\rho(\ell)(\Psi) = \gamma^{\mu}\partial_{\mu}\rho(\ell)(\Psi) + \gamma\rho(\ell)\Psi. \tag{2.5}$$

Equating (2.4) and (2.5) gives the following conditions:

$$[\sigma(\ell)\gamma^{\mu}\sigma(\ell^{-1})][\rho(\ell)\partial_{\mu}\rho(\ell^{-1})] = \gamma^{\mu}\partial_{\mu} \tag{2.6}$$

$$\sigma(\ell)\gamma\sigma(\ell^{-1}) = \gamma. \tag{2.7}$$

(2.7) is readily described: It expresses the fact that γ: $V \to V$ must commute with $\sigma(L)$.

To describe (2.6), notice that the operators

$\rho(\ell)\partial_\mu\rho(\ell^{-1})$ are related to the ∂_μ in the following way (exercise):

$$\rho(\ell)\partial_\mu\rho(\ell^{-1}) = \ell_{\mu\nu}\partial_\nu, \tag{2.8}$$

where $(\ell_{\mu\nu})$ is the 4×4 real matrix such that:

$$(\ell x)_\mu = \ell_{\mu\nu}x_\nu. \tag{2.9}$$

Thus, $\ell \to (\ell_{\mu\nu})$ is the representation of L by (4×4) real matrices which defines L as $O(1, 3)$[1], i.e. the group of 4×4 matrices leaving invariant the Lorentz inner product on R^4, i.e. the inner product with one plus sign and three minus signs. Combining (2.6) with (2.9) gives the following conditions:

---

[1]Recall the definition of $O(p, q)$. Suppose V is a real vector space of dimension $(p + q)$, which is the direct sum of vector spaces $V' + V''$, with: $\dim V' = p$, $\dim V'' = q$. Let β', β'' be symmetric, bilinear, real-valued forms on V' and V''. Suppose that β' is positive definite, β'' is negative definite. Define β as a symmetric, bilinear form on V, as the direct sum of β' and β''. Then, $O(p, q)$ is defined *either* as the group of all automorphisms of V that preserve the form β, *or* as the group of matrices of these automorphisms, defined by choosing a bases of V consisting of orthonormal bases of V' and V'' with respect to β' and β''.

$$\sigma(\ell)\gamma^{\mu}\sigma(\ell^{-1}) = \ell_{\nu\mu}{}^{-1}\gamma^{\nu}. \tag{2.10}$$

(2.7) and (2.8) thus are the conditions that (2.2) be satisfied, i.e. that the operator D commute with $\rho(G)$. They can be described in the language used by physicists by saying that "γ transforms like a scalar", i.e. is invariant under $\sigma(L)$, while "the γ^{μ} transform like a vector", i.e. in the same fashion as the coordinate differential forms dx_{μ} transform. (Of course, if one wants operators $\gamma^{\mu'}$ that transform like the "vector field" $\frac{\partial}{\partial x_{\mu}}$, one has only to use the Lorentz metric tensor g_{μ}, as follows:

$$\gamma^{\mu'} = g_{\mu\nu}\gamma^{\nu}.)$$

In turn, conditions (2.7) and (2.8) can be described at the Lie algebra level. Introduce the basis (J_i, K_i) of the Lie algebra, $\underset{\sim}{L}$, of L, described in Chapter VI, in relations (5.14)-(5.16). σ, as a representation of L, also - by the general principles of Lie group theory - defines a representation $\underset{\sim}{L} \to \sigma(\underset{\sim}{L})$ of $\underset{\sim}{L}$ by operators on V.

Conditions (2.7) and (2.8) now imply that:

$$0 = [J_i, \gamma] = [K_i, \gamma]. \tag{2.11}$$

$$\begin{aligned} &[J_i, \gamma^0] = 0 \\ &[K_i, \gamma^0] = \gamma^i \\ &[J_i, \gamma^j] = -\varepsilon_{ijk}\gamma^k \\ &[K_i, \gamma^j] = -\delta_{ij}\gamma^0. \end{aligned} \tag{2.12}$$

In case V is finite dimensional, conditions (2.12) have been exhaustively investigated by Gel'fand, Minlos and Shapiro [1] and Naimark [1]. Their methods also give results for certain infinite dimensional representations.

Of course, by translating the group conditions (2.7), (2.6), into the "infinitesimal" conditions (2.11), (2.12), we have lost some information; if (2.11) and (2.12) are satisfied there is no guarantee that there is a corresponding representation of the Lorentz group O(1, 3). There are two different things that can go "wrong".

A) O(1, 3) is not connected; the connected component containing the identity is the group $SO^+(1, 3)$, defined as the linear transformation ℓ: $R^4 \to R^4$ such that:

a) determinant $\ell = +1$, i.e. ℓ preserves the "parity" sense.

b) ℓ preserves the "time" sense, i.e. if $x \in R^4$ satisfies: $x_0 > 0$:, then $(\ell x)_0 > 0$ also.

B) $SO^+(3, 1)$ is not simply connected, even though it is connected. In fact, it is known that its simply connected covering group is isomorphic to SL(2, C), the group of 2 × 2 complex matrices of determinant 1.

Phenomena A) and B) turn out to be very important for the physical interpretation of the differential equations one obtains. A) is connected with the question of "parity" and "time reversal" invariance of the theory. B) is related to the "spin" properties of the resulting equations. However, we will not proceed further with these problems at this point.

3. FINITE DIMENSIONAL LINEAR REPRESENTATIONS OF THE LORENTZ LIE ALGEBRA

Let $\underset{\sim}{L}$ be the Lie algebra of the Lorentz group. Recall that we have constructed a basis (J_i, K_i) of $\underset{\sim}{L}$, such that

$$[J_i, J_j] = \varepsilon_{ijk} J_k$$
$$[J_i, K_j] = \varepsilon_{ijk} K_k \qquad (3.1)$$
$$[K_i, K_j] = - \varepsilon_{ijk} J_k$$
$$1 \leq i, j, k \leq 3.$$

Let $\underset{\sim}{L}_c$ be the "complexification of $\underset{\sim}{L}$", i.e. $\underset{\sim}{L}$ is the complex vector space[1] whose basis is the (J_i, K_i), with structure relations (3.1) holding. Then, within $\underset{\sim}{L}_c$ we are free to construct linear combinations of the J's and K's with complex numbers as coefficients. In particular, consider the following elements:

---

[1]In basis-free terms, L_c is the tensor product $L \otimes C$, where C, the complex numbers, is regarded as a real vector space.

$$J_i' = \frac{1}{2}(J_i + \sqrt{-1}\, K_i)$$
$$J_i'' = \frac{1}{2}(J_i - \sqrt{-1}\, K_i). \tag{3.2}$$

Then, the following relations hold:

$$[J_i', J_j''] = 0$$
$$[J_i', J_j'] = \varepsilon_{ijk} J_k' \tag{3.3}$$
$$[J_i'', J_j''] = \varepsilon_{ijk} J_k''$$

To interpret relations (3.3) algebraically, let $\underset{\sim}{L}_u$ be the Lie algebra (a real subalgebra of $\underset{\sim}{L}_c$) generated by the J' and J", and let $\underset{\sim}{K}'$, $\underset{\sim}{K}''$ be the subalgebras of L_u generated by the J' and J". Then;

$$\underset{\sim}{L}_c = \underset{\sim}{L}_u + \sqrt{-1}\, \underset{\sim}{L}_u, \tag{3.4}$$

i.e. $\underset{\sim}{L}_u$ is a real Lie algebra whose "complexification" is isomorphic to the complexification of $\underset{\sim}{L}$. (In terms of the jargon (see "Lie groups for physicists"), $\underset{\sim}{L}_u$ is the "compact real form" of the "non-compact" Lie algebra $\underset{\sim}{L}$.)

$$\begin{aligned} &\underset{\sim}{L}_u = \underset{\sim}{K}' \oplus \underset{\sim}{K}'' \\ &[\underset{\sim}{L}_u, \underset{\sim}{K}'] \subset \underset{\sim}{K}' \\ &[\underset{\sim}{L}_u, \underset{\sim}{K}''] \subset \underset{\sim}{K}'' \\ &[\underset{\sim}{K}', \underset{\sim}{K}''] = 0. \end{aligned} \tag{3.5}$$

Relations (3.5) show that $\underset{\sim}{L}_u$ is the direct sum of the two ideals; $\underset{\sim}{K}'$, $\underset{\sim}{K}''$, and each of these ideals is a Lie algebra isomorphic to the Lie algebra of SO(3, R) (or SU(2)).

Thus, linear representations of $\underset{\sim}{L}_u$ can be constructed in the following way: Let $\rho_1(\underset{\sim}{K}')$, $\rho_2(\underset{\sim}{K}'')$ be representations of $\underset{\sim}{K}'$ and $\underset{\sim}{K}''$ by operators on vector spaces V_1, V_2. Form:

$$V = V_1 \otimes V_2, \tag{3.6}$$

i.e. V is the "tensor product" of V_1 and V_2. Define a representation ρ of $\underset{\sim}{L}_u$, as follows:

$$\begin{aligned} &\rho(X' + X'')(v_1 \otimes v_2) \\ &= \rho_1(X')(v_1) \otimes v_2 + v_1 \otimes \rho_2(X'')(v_2) \end{aligned} \tag{3.7}$$

for $X' \in K'$, $X'' \in K''$.

Exercise: Show that this is indeed a representation of $\underset{\sim}{L}_u$. Show also that every irreducible representation of $\underset{\sim}{L}_u$ be operators on a complex finite dimensional vector space V is of this form, with ρ_1, ρ_2 irreducible. Show conversely, that ρ is irreducible if ρ_1, ρ_2 are irreducible.

Thus, we can construct representations of $\underset{\sim}{L}_u$ if we know representations of the Lie algebra of SO(3, R). In turn, this will provide us with representations of $\underset{\sim}{L}$ *providing the vector space is complex*, since the operators of $\underset{\sim}{L}$ can be described as linear combination of those of $\underset{\sim}{L}_u$ with complex numbers. (The analysis of which of these representations of $\underset{\sim}{L}$ are "real" is more delicate; we will put it off until it is needed.)

Then, we can reduce the search for finite dimensional representations of $\underset{\sim}{L}$ to the finite dimensional representations of $\underset{\sim}{K}$, the Lie algebra of SO(3, R). This analysis is well-known, (for example, see "Lie algebra for physicists", Wigner [1], or any quantum mechanics text, for that matter; e.g., Gottfried [1]) hence we will only state the result without proof.

<u>THEOREM 3.1</u>. Let $\underset{\sim}{K}$ be the Lie algebra generated by the (J_i), and let $\rho(\underset{\sim}{K})$ be an irreducible representation of $\underset{\sim}{K}$ by operators on a finite dimensional vector space V. Then, there is an integer or half-integer "s" (called the *spin* of the representation) characterizing the representation, in the sense that V has a basis (v_m), $-s \leq m \leqq s$, such that:

$$\begin{aligned} &\text{a)} \quad \rho(J_3)(v_m) = imv_m \\ &\text{b)} \quad \rho(J_+)(v_m) = \sqrt{s(s+1)-m(m+1)}\; v_{m+1} \qquad (3.8) \\ &\text{c)} \quad \rho(J_-)(v_{m+1}) = \sqrt{s(s+1)-m(m+1)}\; v_m. \end{aligned}$$

The notations used in (3.8) are as follows:

There is no summation on m. Thus, a) expresses the fact that v_m is an eigenvector, with eigenvalue m, of the "Hermitian" operator

$$- i\rho(J_3).$$

In the Dirac notations,

$$v_m = |m>$$

J_+ and J_- are defined as follows:

$$\begin{aligned} J_+ &= J_2 - iJ_1 \\ J_- &= -iJ_1 - J_2. \end{aligned} \tag{3.9}$$

Exercise: Show that the integer values of s give representations of K = SO(3, R). Show that this is not so for the half-integer values, but the representations (3.8) arise from a faithful[1] representation of the simply connected covering group K' = SU(2).

The facts about tensor products of representations of $\underset{\sim}{K}$ are equally well-known, hence will be presented without proof:

THEOREM 3.2. Suppose $\rho_{s_1}(\underset{\sim}{K})$, $\rho_{s_2}(\underset{\sim}{K})$ are two irreducible representations of $\underset{\sim}{K}$ on vector spaces V_1, V_2, of spins s_1, s_2. Let $\rho(\underset{\sim}{K})$ be the "tensor product":

---

[1]A representation is said to be *faithful* if it only maps the identity into the identity operators, i.e., if its kernel is the identity.

$$\rho(X)(v_1 \otimes v_2) = \rho_1(X)v_1 \otimes v_2$$
$$+ v_1 \otimes \rho_2(X)(v_2) \tag{3.10}$$
$$\text{for} \quad X \in K,\ v_1 \in V_1,\ v_2 \in V_2.$$

Then, $V_1 \otimes V_2$ is the direct sum of subspaces

$$V^{|s_1-s_2|},\ V^{|s_1-s_2+|}, \ldots,\ V^{s_1+s_2},$$

in each of which $\rho(\underset{\sim}{K})$ acts irreducibly-acting as the spin s representation in V^s. Somewhat symbolically then, we have the "Clebsch-Gordan" series:

$$\rho_{s_1} \otimes 1 + 1 \otimes \rho_{s_2} = \sum_{s=|s_1-s_2|}^{s_1+s_2} \rho_s. \tag{3.11}$$

Let us return to the consideration of $\underset{\sim}{L}_u$, $\underset{\sim}{L}_c$ and $\underset{\sim}{L}$, $\underset{\sim}{K}'$ and $\underset{\sim}{K}''$. As we have seen, the irreducible representations of $\underset{\sim}{L}$ on complex vector spaces can be represented as tensor products of representations of $\underset{\sim}{K}'$ and $\underset{\sim}{K}''$, which are both isomorphic to $\underset{\sim}{K}$. Hence, the irreducible representations of $\underset{\sim}{L}$

can be labelled as (s_1, s_2), where s_1 and s_2 refer to the spin of the representation of $\underset{\sim}{K}'$ and $\underset{\sim}{K}''$ from which the representation of $\underset{\sim}{L}$ or $\underset{\sim}{L}_u$ is built up by the tensor product. The relation to keep in mind is that the dimension of the vector space of the representation labelled (s_1, s_2) is:

$$(2s_1 + 1)(2s_2 + 1). \tag{3.12}$$

The Clebsch-Gordan series (3.11) plays a key role in this description of the representations of the Lorentz group. Suppose $\rho(\underset{\sim}{L}_u)$ is described as a tensor product.

$$\rho(\underset{\sim}{K}' + \underset{\sim}{K}'') = \rho_1(\underset{\sim}{K}') \otimes 1 + 1 \otimes \rho_2(\underset{\sim}{K}''). \tag{3.13}$$

Let $\underset{\sim}{K}$ be the subalgebra of $\underset{\sim}{L}_u$ and $\underset{\sim}{L}$ generated by the J_i. (In terms of the jargon, it is a *maximal compact subalgebra* of the noncompact semisimple Lie algebra $\underset{\sim}{L}$.) From (3.2), we have:

$$J_i = J_i' + J_i'' \tag{3.14}$$

Thus, $\underset{\sim}{K}$, as a subalgebra of $\underset{\sim}{L}_u = \underset{\sim}{K}' \oplus \underset{\sim}{K}''$ is identified with the "diagonal" subalgebra of $\underset{\sim}{K}' + \underset{\sim}{K}'' = \underset{\sim}{K} + \underset{\sim}{K}$. In particular, we see that, with ρ defined by (3.13), $\rho(K)$ will decompose into irreducible representations according to the Clebsch-Gordan series (3.11). For example, this leads us to several remarks about the identification of representations of $\underset{\sim}{L}$ described in this "spinorial" way with representations described more "physically" or "geometrically". First, some exercises will provide us with certain needed data, and will provide the reader with some needed practice with the formalism. (It is not necessary, however, to do all of them.)

<u>Exercise</u>: If the simply connected covering group of $L = SO^+(1, 3)$ is identified with $SL(2, C)$, identify the generators J_i', J_i'' with 2×2 complex matrices, i.e. elements of the Lie algebra of $SL(2, C)$.

<u>Exercise</u>: Show that the connected subgroup of $SL(2, C)$ corresponding to the subalgebra $\underset{\sim}{K}$ is the subgroup $SU(2)$ of 2×2 unitary matrices of determinant one.

<u>Exercise</u>: Show that the representation $\underset{\sim}{L}$ labelled $(\frac{1}{2}, 0)$ corresponds to the "defining" representation of SL(2, C) by 2 × 2 matrices. Show that $(0, \frac{1}{2})$ is the "complex-conjugate" representation. (Physicists often call the vector spaces corresponding to $(\frac{1}{2}, 0)$ and $(0, \frac{1}{2})$ "undotted" and "dotted" spinors.)

<u>Remark</u>. The result of this exercise is compatible with (also suggested by, and, indeed, probably provable from) the Clebsch-Gordan result. If $\rho = (\frac{1}{2}, 0)$, by the above remarks, $\rho(\underset{\sim}{K})$ remains irreducible, giving the spin $\frac{1}{2}$ representation.

<u>Exercise</u>: Show that the $(\frac{1}{2}, \frac{1}{2})$ representation of $\underset{\sim}{L}$ is the 4-dimensional "real" representation which is the "defining" representation of $L = SO^+(3, 1)$. Precisely, if $\rho(\underset{\sim}{L})$ acts on the complex four dimensional vector space V, show that $\rho(\underset{\sim}{L})$ leaves invariant a four *real* dimensional subspace V_0, and that V_0 has on it a symmetric bilinear form whose canonical form has one "plus", and three "minus" signs. Also, prove it the other way: Starting off with $\rho_0(\underset{\sim}{L})$ acting on R^4 via the defining repre-

sentation $SO^+(4, 1)$, extend $\rho_0(\underset{\sim}{L})$ to $\rho(\underset{\sim}{L})$ acting on C^4 by "complexifying", and show that this gives the $(\frac{1}{2}, \frac{1}{2})$ representation.

Remark. Again, Clebsch-Gordan provides a way of guessing this result. If $\rho = (\frac{1}{2}, \frac{1}{2})$, then $\rho(K)$ reduces to $0 \oplus 1$, which is compatible with the result of the last exercise, since the "vector" representation of $O(3, 1)$ splits under $O(3, R)$ into a "scalar" and "vector" representation of $O(3, R)$.

Exercise: Show that (s_1, s_2) is equivalent to the tensor product

$$(s_1, 0) \otimes (0, s_2).$$

In physicist's language this expresses a general "spinor" in terms of "undotted" and "dotted" spinors.

Exercise: Show that the kernel of the covering-group homomorphism: $SL(2, C) \to L = SO^+(1, 3)$ is Z_2, the abelian group with two elements; say, g_0, g_1, where g_0 is the identity element, and g_1

is a nonidentity element of SL(2, C). Show that g_1 is in the center of SL(2, C).
(Hint: First determine the center of SL(2, C).)
Then, express g_0 in terms of the generators J_i, J_i', J_i''.

Exercise: Let ρ be the (s_1, s_2) representation of $\underset{\sim}{L}$. From general Lie group principles, ρ arises from a representation of SL(2, C). Calculate $\rho(g_0)$, and thus determine whether or not ρ arises also from a representation of $SO^+(1, 3)$.

Exercise: Let L be O(1, 3), and let $L_0 = SO^+(1, 3)$ be its maximal connected subgroup. From general Lie group principles, L_0 is an invariant subgroup of L.

a) Show that L/L_0 is a finite group, isomorphis to $Z_2 \times Z_2$, i.e.

b) Show that L has a subgroup S such that:

$$L = S \cdot L_0,$$

i.e. L is a semidirect product of S and L, with L/L_0 isomorphic to S.

c) Identify elements, labelled ℓ_P, ℓ_T of S such that:

$$\ell_P^{\,2} = 1 = \ell_T^{\,2};\ \ell_P \ell_T = \ell_T \ell_P,$$

i.e. S is generated by 1, ℓ_P, ℓ_T, $\ell_P \ell_T$. As the labels "P", "T" indicate, ℓ_P, ℓ_T should be chosen so that they correspond, physically, to "parity" and "time reversal".

d) Let $\ell_P = \text{Ad } \ell_P$, $\ell_T = \text{Ad } \ell_T$, be the corresponding automorphisms of L_0 and L. Calculate the effect of ℓ_P and ℓ_T on the (J, K)-basis for $\underset{\sim}{L}$.

d) If $\rho(L_0)$ is a representation of L_0, leading to a representation labelled (s_1, s_2) in the "spinorial" notation, for which values of s_1 and s_2 can this representation be extended to L? Is such an extension unique, if it exists?

f) Suppose $\rho(\underset{\sim}{L})$ is a representation of the Lie algebra, labelled (s_1, s_2) in the spinorial notation. For which values of (s_1, s_2) does there exist a representation

$\rho'(S)$ of the group S, on the same vector space as ρ, such that:

$$\rho(\text{Ad } g(X)) = \rho'(g)\rho(X)\rho'(g^{-1}) \quad (3.15)$$

for all $g \in S$, all $X \in \underset{\sim}{L}$.

<u>Exercise</u>: Here is another - more group theoretic approach - to some of the problems sketched in the last exercise.

Let P, T: $\underset{\sim}{L} \to \underset{\sim}{L}$ be the automorphisms of the Lorentz Lie algebra which were denoted by ℓ_P, ℓ_T in the last exercise, and which physically correspond to "parity" and "time reversal".

Let L be the simply connected group whose Lie algebra is $\underset{\sim}{L}$. By general principles of Lie group theory, P, T arise from automorphism - which we will denote by the same letter - of L.

a) In terms of identification of L with SL(2, C), write down P and T explicitly.

b) Let S be the group of automorphisms of SL(2, C) generated by the automorphism P, T constructed in part a). Write down the "structure" of S, i.e. write down the relations satisfied by P, T and their products. Let L' be the semidirect product of S·L.

c) If $\rho = (s_1, s_2)$ is an irreducible representation of L, what are the conditions on (s_1, s_2) that guarantee that ρ can be extended to a representation of L' on the same vector space.

d) More generally, classify the irreducible, finite dimensional representation of L'. Are all finite dimensional representations of this group completely irreducible, i.e. equivalent to a direct sum of irreducibles?

The following exercises require more algebraic expertise on the part of the reader - indeed, they may not even be in the literature - but they probably may be regarded as equivalent to "folklore" among the physicists who work on "discrete symmetries".

First a definition:

DEFINITION. Let G', G be groups. A homomorphism ϕ: G' → G which is onto G (i.e. a representation of G as a quotient of G by the invariant subgroup H, the kernel of ϕ), is said to define G' as an *extension* of G by H.

The extension is said to be *abelian* of H is abelian. It is said to be a *central* extension if H is in the center of G', i.e. each $h \in H$ commutes with every element of G'.

The central extensions play a key role in the determination of the "ray representations" of the group G. (See Wigner [1] and Bargmann [1]).

Exercise: If L' is the group S·L described in the last exercise, determine all central extensions G' of L', where G' is a Lie group, with the homomorphism ϕ: $G' \to L'$ defining the extension Lie group homomorphism.

Exercise: For all groups G' found in the last exercise (or at least all those that the reader succeeds in constructing, if he is not able to find them all) classify all irreducible representations, with particular emphasis on the properties of these representations vis-a-vis the projection onto L'. Discuss in broad qualitative terms the relevance of all this to the classification of the "discrete symmetries" of elementary particles. (For related material with a different emphasis, see Hermann [6], Lee and Wick [1], Zumino and Zwanziger [1].)

Finally, the following exercise provides a useful formula, and only involves the Clebsch-Gordan series for SU(2).

<u>Exercise</u>: Suppose $\rho_1 = (s_1, s_2)$, $\rho' = (s_1', s_2')$ are two irreducible representations of $\underset{\sim}{L}$. Let $\rho = \rho_1 \otimes 1 + 1 \otimes \rho_2$, i.e. ρ is the "tensor-product" representation (in "infinitesimal form"). Show that:

$$\rho = \sum_{\substack{|s_1-s_1'| \leq j \leq s_1+s_1' \\ |s_2-s_2'| \leq k \leq s_2+s_2'}} (j, k) \qquad (3.16)$$

(The "equality sign" for representations means "equivalence", in the sense that there exists an intertwining operator which establishes an isomorphism between the corresponding vector spaces.) As we will show in the next section, this result is useful for carrying out the analysis of the existence of first-order invariant differential operators.

4. CONDITIONS FOR INVARIANT FIRST ORDER DIFFERENTIAL OPERATORS ON POINCARÉ-HOMOGENEOUS VECTOR BUNDLES

Let us recapitulate. Let L = O(1, 3) be the Lorentz group, and G, the Poincaré group, be a semidirect product L·T of L with the four dimensional translation group. G acts as a transformation group on R^4, i.e., L is the "geometric" Poincaré group of automorphisms of space-time that preserve the Lorentz metric.

Let L' be a group, with ϕ: L' → L a homomorphism of L' into L. This defines a homomorphism of L' into the group of isomorphisms of T: First map $\ell' \in L'$ into $\ell = \ell(\ell')$, then assign to ℓ the automorphism $t \to \ell t \ell^{-1}$ of T. This in turn enables us to construct the semidirect product

$$G' = L' \cdot T.$$

The homomorphism ϕ: L' → L then extends to a

homomorphism: $G' \to G$, with G mapped onto itself via the identity map. In turn, this enables one to consider G' as a transformation group on R^4.

Let E be a vector bundle over R^4 that is acted on homogeneously by G'. By the work of Section 1, E is isomorphic to the product: $R^4 \times E(0)$, and everything is determined by the isotropy representation of L' on $E(0) = V$, denoted by σ. In particular, a first-order, invariant (under the action of G') linear differential operator: $\Gamma(E) \to \Gamma(E)$ is determined by five linear maps γ^μ, γ: $V \to V$ such that:

a) γ commutes with $\sigma(L')$

b) The γ^μ transform "like a four-vector" under $\sigma(L')$. (4.1)

This is not precisely the context with which we worked in sections 1 and 2. (There, G was equal to G'). However, it is readily seen that the work done there generalizes to prove (4.1). There are several interesting situations in which this generalization is useful.

$L' = SL(2, C)$, the simply connected covering

group of $SO^+(1, 3)$. Then the representation σ can be described (if V is finite dimensional, which is the situation with which we are working here) in the spinorial terms and notation (s_1, s_2)

$$L' = SL(2, C) \times S, \tag{4.3}$$

where S is a compact, connected Lie group, and ϕ: $L' \to L$ is the projection on SL(2, C), followed by the covering map: $SL(2, C) \to SO^+(1, 3)$.

In case (4.3), σ can be constructed out of tensor products of the spinorial representations of SL(2, C) and the irreducible representation of the group S. In physical language, S is the "internal symmetry group". Say S = SU(2) or SU(3), for typical isotopic spin or Eightfold Way theories of elementary particles.

Our main problem in this chapter is to analyze condition (4.1b) in more precise terms. To do this let us present some general algebraic remarks.

Let H be a Lie group, and let $\rho(H)$ be a linear representation of H by operators on a complex vector space V. Let L(V) be the Lie algebra (under commutator) of linear maps A: $V \to V$. ρ defines a representation ρ' of H by linear transformations on L(V) as follows:

$$\rho'(h)(A) = \rho(h)A\rho(h^{-1}) \tag{4.4}$$

A problem that is frequently encountered in physical problems - for example, in analyzing (4.1) - is to decompose $\rho'(H)$ into irreducible representations. In case H is the Lorentz group, the following further general algebraic remarks are useful.

A dual vector to V is a complex-linear map θ: $V \to C$. The collection of these vectors forms a complex vector space, called the *dual space* to V, and denoted by: V^d : The dimension of V^d is equal to the dimension of V (exercise).

If A: $V \to V$ is a linear transformation, there is a linear transformation, denoted by A^d: $V \to V$, defined as follows:

$$A^d(\theta)(v) = \theta(Av) \tag{4.5}$$

$$\text{for} \quad \theta \in V^d, \ v \in V$$

This correspondence $A \to A^d$ satisfies the following rule:

$$(A_1A_2)^d = A_2{}^d A_1{}^d \qquad (4.6)$$

for $A_1, A_2 \in L(V)$.

If $\rho(H)$ is a linear representation of the group H by operators on V, there is a "dual" representation - denoted by ρ^d - of H by operators on V^d:

$$\rho^d(h) = \rho(h^{-1})^d \qquad (4.7)$$

At the Lie algebra level, denote by $\rho(X)$, for $X \in H$, the following linear transformation of V:

$$\rho(X) = \frac{d}{dt}\rho(\exp(tX))|_{t=0}, \qquad (4.8)$$

i.e. $\rho(H)$ is the "infinitesimal version" of the representation $\rho(H)$. From (4.8) and (4.7), one has the following relation:

$$\rho^d(X) = -\rho(X)^d \qquad (4.9)$$

for $X \in H$

Also, (4.4) and (4.8),

$$\rho'(X)(A) = [\rho(X), A]$$

for $A \in L(V)$, $X \in H$

Let V" denote the tensor product:

$$V'' = V \otimes V^d \tag{4.10}$$

Define:

$$\rho''(h)(v \otimes \theta) = \rho(h)(v) \otimes \rho^d(h)(\theta)$$

for $h \in H$, $v \in V$, $\theta \in V^d$

i.e. ρ'' is the "tensor product" $\rho \otimes \rho^d$.

From (4.8) and (4.11), we have, at the Lie algebra level,

$$\rho''(X)(v \otimes \theta) = \rho(X)(v) \otimes \theta + v \otimes \rho^d(X)(\theta)$$

$$= \rho(X)(v) \otimes \theta - v \otimes \rho(X)^d(\theta) \tag{4.12}$$

Symbolically,

$$\rho''(H) = \rho(H) \otimes 1 + 1 \otimes \rho^d(H)$$

Now, define a linear map α: $V \otimes V^d \to L(V)$, as follows:

$$\alpha(v_1 \otimes \theta)(v) = \theta(v)v, \tag{4.13}$$

for $v, v_1 \in V, \theta \in V^d$

We now prove several properties of α:

<u>THEOREM 4.1</u>. α is an isomorphism between the vector spaces $V'' = V \otimes V^d$ and $L(V)$.

<u>Proof</u>. Suppose that (v_j) $1 \leq i, j \leq n =$ dim V, is a basis of V. Let (θ_i) be the "dual basis" of V^d, i.e.

$$\theta_i(v_j) = \delta_{ij}$$

Suppose that $v'' \in V''$ is of the form:

$$v'' = a_{ij} v_i \otimes \theta_j, \tag{4.14}$$

with $\alpha(v'') = 0$.

Now, using (4.13), we have:

$\alpha(v'')(v_k)$

$= a_{ij}(v_i \times \theta_j)(v_k) = a_{ij}\delta_{jk}v_i$

$= a_{ik}v_i$, whence: $a_{ik} = 0$: (4.15)

This shows that α is an isomorphism, since it is a one-one map between vector spaces of the same dimension. This calculation also shows that the "tensor" coefficient (a_{ij}) determining v" are also the matrix of the linear transformation $\alpha(v'')$.

<u>THEOREM 4.2</u>. α intertwines the action of H and $\underset{\sim}{H}$ on V" and L(V) described above.

<u>Proof</u>. Use (4.13). For $h \in H$, v, $v \in V$, $\theta \in V^d$,

$\rho'(h)(\alpha(v_1 \otimes \theta))(v)$

$= \rho(h)\alpha(v_1 \otimes \theta)\rho(h^{-1})(v)$

$= \rho(h)(\theta(\rho(h^{-1}v))v_1)$

$= \rho^d(h)(\theta)(v)\rho(h)(v_1)$

$$= \alpha(\rho(h)(v_1) 0 \; \rho^d(h)(\theta))$$

$$= \alpha(\rho''(h)(v_1 \times \theta))$$

which is the intertwining property.

The property for H is proved similarly.

<u>THEOREM 4.3</u>. The representations ρ and ρ^d are equivalent, i.e., ρ is *self-dual*, if an only if there is complex-valued bilinear form $\beta: V \times V \to C$ which is invariant under the action of $\rho(H)$ and which is non-degenerate.

<u>Proof</u>. If ρ and ρ^d are equivalent, there is an H-intertwining isomorphism α: $V \to V^d$. Define $\beta = V \times V \to C$ as follows:

$$\beta(v_1, v_2) = \alpha(v_2)(v_1) \tag{4.15}$$

for $v_1, v_2 \in V$.

Then, for $h \in H$,

$$\beta(\rho(h)v_1, \rho(h)v_2) = \alpha(\rho(h)(v_2))(\rho(h)(v_1))$$

$$= \rho^d(h)\alpha(v_2)(\rho(h)(v_1))$$

$$= \alpha(v_2)(\rho(h^{-1})\rho(h)(v_1))$$

$$= \alpha(v_2)(v_1) = \beta(v_1, v_2) \tag{4.16}$$

(4.16) now expresses the fact that β is an invariant bilinear form. That β is non-degenerate (i.e. that $\beta(v, V) = 0$ implies $v = 0$) and that the existence of β such a β implies self-duality is left to the reader to prove.

<u>THEOREM 4.4</u>. Suppose $H = H_1 \times H_2$, i.e. H is the direct product of invariant subgroups H_1, H_2. Let $\rho_1(H_1)$, $\rho_2(H_2)$ be representations of H_1 and H_2 on vector spaces V_1, V_2, and construct:

$$V = V_1 \otimes V_2$$

$$\rho(H)\colon \rho(h_1, h_2) = \rho_1(h_1) \otimes \rho_2(h_2) \tag{4.17}$$

$$\text{for} \quad h_1 \in H_1,\ h_2 \in H_2$$

Then if ρ_1 and ρ_2 are self-dual, so is the tensor product ρ.

<u>Proof</u>. Let $\beta_1\colon V_1 \otimes V_2 \to C$, $\beta_2\colon V_2 \otimes V_2 \to C$

be the non-degenerate forms provided by the hypothesis, and Theorem 4.3. Let β be the "tensor product" form $\beta_1 \otimes \beta_2$, defined as follows:

$$\beta(v_1 \otimes v_2, v_1', v_2') = \beta_1(v_1, v_1')\beta_2(v_2, v_2') \quad \text{for} \quad v_1, v_1' \in V_1, v_2, v_2' \in V_2 \tag{4.18}$$

Then, β defined by (4.18) is non-degenerate (exercise), and is invariant under $\rho(H)$ hence, by Theorem 4.3 again, ρ is self-dual.

Remark. Notice from (4.18) that the following facts hold. They are useful in a variety of places in elementary particle physics. If β_1, β_2 are *both* symmetric *or* skew symmetric, then $\beta = \beta_1 \otimes \beta_2$ is symmetric. If one of the forms β_1, β_2 are symmetric, the other skew-symmetric, then β is skew-symmetric.

Now, several exercises further along those lines will be useful to the reader as practice in the formalism, and to the later development of the theory.

Exercise: If $\rho(H)$ is irreducible, and is self-dual, then the form β is either symmetric or skew-symmetric. (4.19)

Exercise: Each irreducible representation of $H = SU(2)$ is self-dual. The form β is symmetric for those of integer spin; skew-symmetric for those of half-integer spin. (4.20)

Now let us return to the situation described in the beginning of the section. Let $H = L' = SL(2, C)$, and suppose $\rho(L')$ is a representation of L' on a vector space V which is the fibre over the point $0 \in R^4$ of a G' homogeneous vector bundle over R^4. We are concerned with the existence of operators (γ^μ), $(\gamma) \in L(V)$ that transform like the "vector" representation, i.e. the representations (1/2, 1/2) and (0, 0) in the spinorial notation for the representations of $SL(2, C)$.

Exercise: The representation (s_1, s_2) of $SL(2, C)$ is self-dual i.e. is (s_1, s_2). (4.21)

With (4.21), we are prepared for serious work. Our problem is then reduced - using Theorem 4.1 and 4.2 - to the reduction of

$$\rho(L) \otimes \rho^{d}(L)$$

into irreducible representations, and counting how many times the (1/2, 1/2) and the (0, 0) representation occur.

Suppose, for example, that:

$$\rho = \sum_{(s_1,\, s_2)\varepsilon I} (s_1,\, s_2) \tag{4.22}$$

where I denotes some set of ordered pairs of integers or half-integers (possibly repeated). Thus (4.22) means the ρ when reduced into the direct sum of irreducible representations, is the direct sum of the representations which - in the spinorial notation - occur on the right hand side of (4.16). Using (4.21), we then have

$$\rho^{d} = \sum_{(s_1,\, s_2)\varepsilon I} (s_1,\, s_2)$$

$$\rho \otimes \rho^{d} = \sum_{\substack{(s_1,\, s_2)\varepsilon I \\ (s_1',\, s_2')\varepsilon I}} (s_1,\, s_2) \otimes (s_1',\, s_2') \tag{4.23}$$

In turn, this can be worked out using (3.16). Rather than worry about the general case, we shall do some examples.

EXAMPLE 1.

$$\rho = (s_1, s_2)$$

Then, using (4.21) and (4.16),

$$\rho \otimes \rho^d = (s_1, s_2) \otimes (s_1, s_2)$$

$$= \sum_{0 \leq j,\ k < s_1 + s_2} (j, k) \tag{4.24}$$

Since in the sum (4.24), the sum must go by integer steps, (1/2, 1/2) cannot appear, i.e. there is no Lorentz invariant wave equation.

EXAMPLE 2. The Dirac Equation

$$\rho = (1/2, 0) + (0, 1/2)$$

Then,

$$\rho^d = \rho$$

$$\rho \otimes \rho^d = [(1/2,\ 0) + (0,\ 1/2)]$$

$$\otimes [(1/2,\ 0) + (0,\ 1/2)] = (1/2,\ 0)$$

$$\otimes (1/2,\ 0) + (0,\ 1/2) \otimes (1/2,\ 0)$$

$$+ (1/2,\ 0) \otimes (0,\ 1/2) + (0,\ 1/2)$$

$$\otimes (0,\ 1/2) = (0,\ 0) + (1,\ 0)$$

$$+ (1/2,\ 1/2) + (1/2,\ 1/2) + (0,\ 0)$$

$$+ (0,\ 1) \qquad (4.25)$$

Thus, we see that indeed L(V) has a set of elements transforming like a "four vector". Indeed, it has two such sets. When additional physical conditions of invariance under parity and time-reversal are imposed, this ambiguity is resolved, and there is a unique such set of operators. The resulting differential operator:

$$D = \gamma^\mu \partial_\mu + i\ m: \qquad (4.26)$$

(with m is a real number) is called the *Dirac operator*. The corresponding differential equation: D = 0: is called the *Dirac equation*.

We now turn to the study of the more profound reason why this equation is useful - namely, that it "describes" free particles of "mass" m and "spin" 1/2.

5. MASS AND SPIN RELATIONS FOR POINCARÉ-INVARIANT DIFFERENTIAL OPERATORS

Let $G = L \cdot T$ be the "geometric" Poincaré group, the semidirect product of the Lorentz group $L = O(1, 3)$, and the four dimensional translation group: $T = R^4$:. Let L' be the simply connected group $SL(2, C)$, covering the connected group $SO^+(1, 3)$, and let $G' = L' \cdot T$, the semidirect product of L' and T, be the "physical" Poincaré group. G' also acts on R^4 via its homomorphism: $G' \to G$ (although not "faithfully", of course). Let $E \to R^4$ be a G'-homogeneous vector bundle over R^4, let $\rho(G')$ be the representation of G' in $\Gamma(E)$ obtained by letting G' act on the cross-sections in the usual way, and let $D: \Gamma(E) \to \Gamma(E)$ be a differential operator that commutes with the action of $\rho(G')$.

Let $\Gamma(E, D)$ be the space of $\Psi \in \Gamma(E)$ such that:

$$D(\Psi) = 0 \tag{5.1}$$

Then, $\rho(G')$ maps $\Gamma(E, D)$ onto itself. Our problem on this section is to apply the general machinery of Chapter VI, and to decompose this representation of G' into irreducible representations.

Let $V = E(0)$ be the fiber of E over the point "zero" of R^4. Then, $\Gamma(E)$ can be identified - as explained in Section 1 - with the space of maps $\Psi\colon\ R^4 \to V$. Let $\sigma(L')$ be the isotropy representation of L' on V. Then,

$$\rho(\ell')(\Psi)(x) = \sigma(\ell')\Psi(\ell^{-1}x)$$

$$\text{for} \quad \ell' \in L', \ x \in R^4,$$

with ℓ the image of $\ell' \in L'$ in $L = SO^+(1, 3)$ under the covering map. It is of the following form:

$$D(\Psi) = \gamma^\mu \partial_\mu \Psi + \gamma\Psi \tag{5.2}$$

where γ^μ, γ are linear operators: $V \to V$. The γ^μ transform under $\rho(L')$ "like a 4-vector", γ transforms "like a scalar".

Identify T with the collection of $\{g_a : a\ R^4\}$, with: $g_a(x) = x + a$. Then T^d is identified with the collection of linear maps λ_p: $T \to C$ of the form:

$$\lambda_p(g_a) = e^{-ip\cdot a} \tag{5.3}$$

where $p = (p_\mu) \in C^4$ is a *complex*[1] four vector.

If $\lambda_p \in T^d$ of form (5.3) is to be an eigenvalue of $\rho(T)$, obviously the eigenvector $\Psi_p \in \Gamma(E)$ must be of the following form:

$$\Psi_p(x) = e^{ip\cdot x}v \tag{5.4}$$

where v is some element of V. However, v is restricted by the further condition that $\Psi_p \in \Gamma(E, D)$, since then; by (5.1), (5.2),

---

[1]Notice that we are now leaving open the possibility that some non-unitary representations of G' will appear in the decomposition of ρ. Notice also that in this section p_μ denotes the *components* of the four vector p_μ, and *not* a basis of R^4, as was the case in previous work.

$$0 = ip_\mu \gamma^\mu(v) + \gamma(v) \quad (5.5)$$

Now, for $p \in C^4$, let $V(p)$ be the space of all vectors $v \in V$ which satisfy (5.5). Thus, $V(p)$ can be identified with what in Chapter 4 we denoted by $\Gamma(E, D)^{\lambda_p}$, i.e. the space of eigenvectors of $\rho(T)$ that lie in $\Gamma(E, D)$.

Let M denote the space of all $p \in C^4$ such that $V(p) \neq (0)$. Construct a "vector bundle"[1] $E' \to M$ as follows: E' is the subset $C^4 \times V$ consisting of the pairs $(p, v) \in C^4 \times V$ such that:

$$p \in M; \; v \in V(p) \quad (5.6)$$

Thus, the fiber $E'(p)$ is identified with $V(p)$, and we will so denote it. G' acts on E' as follows:

$$g_a(p, v) = (p, e^{ia \cdot p} v) \quad (5.7)$$

for $a \in R^4$, $p \in M$, $v \in V(p)$

[1] We put vector bundle in quotes since it is not necessarily a vector bundle in the local-product sense, but may be some more general object. However, those objects for simple equations like the Dirac equation are local product, so what we say will have a literal sense for those cases.

$$\ell'(p, v) = (\ell p, \sigma(\ell')(v)), \tag{5.8}$$

for $p \in M$, $v \in V(p)$, $\ell' \in L$,

where ℓ is the image of $\ell' \in SL(2, C)$ in $SO^+(1, 3)$.

Notice that in order to prove that ℓ' defined by (5.8) actually maps E' into E', it is necessary to show that:

$$\sigma(\ell')(V(p)) = V(\ell p) \tag{5.9}$$

Exercise. Prove (5.9), from the relations (5.5) defining V(p) and from the "vectorial" and "scalar" transformation law of the γ^μ and γ under $\sigma(L')$.

Thus, we can replace the problem of decomposing $\rho(G')$ acting in $\Gamma(E, C)$ to the problem of decomposing the action of G' on the cross-sections $\Gamma(E')$. Recall from Chapter VI that there are two major problems here:

A) Find the orbits of $L = SO^+(1, 3)$ acting on M

B) On each of the orbits found in A), pick a point p^o and decompose the linear isotropy representation of the isotropy subgroup L'^{p^o} on the fiber $V(p^o)$.

In turn, this suggests that we calculate $V(p^o)$ and $\sigma(L'^{p^o})$ for three families of points of C^4.

$$p^o = (m, 0, 0, 0), \tag{5.10}$$

with $m \in C$

$$p^o = (0, m, 0, 0), \tag{5.11}$$

with $m \in C$

$$p^o = (1, 1, 0, 0) \tag{5.12}$$

Let us first work with (5.10). From (5.5), $V(p^o)$ then consists of the $v \in V$ such that:

$$\gamma^o(v) = \frac{i}{m}\gamma(v)$$

L'^{p^o} is the SU(2)-subgroup of $L' = SL(2, C)$ that we will denote by K.

Now, let us assume[1] that V can be written as

[1] Of course, if V is finite dimensional, these assumptions can be proved. They can also be proved for certain types of infinite dimensional "representations" (See Naimark [1]. We use quotes for representations because we want to allow more general possibilities than Naimark of representations that are not global representations of the group, but may have singularities.

the direct sum of subspaces labelled $\{V^j\}$, where j runs through a set of integers or half-integers, with $\sigma(K)$ mapping each V^j onto itself, and acting within V^j as a direct sum of a number of copies of the spin j - irreducible representation of $K = SU(2)$.

Now, the operator $\gamma^o - \frac{i}{m}\gamma$: $V \to V$ commutes with $\sigma(K)$, hence will map V^j onto itself. Thus, we have proved the following criterion:

THEOREM 5.1. A representation of G' of "mass" m (possibly complex) and "spin" j will occur in the decomposition of $\sigma(G')$ on $\Gamma(E, D)$ if the operator

$$\gamma^o - \frac{i}{m}\gamma$$

acting in V^j is not one-one.

The analysis of cases (5.11) and (5.12) is similar, but complicated by the fact that the isotropy groups are non-compact, hence one must envisage more complicated decomposition schemes for V. Let H be the isotropy subgroup of L' at p^o, in cases (5.11), (5.12). Suppose the irreducible representations of H are labelled by a complex parameter j'. Let $V^{j'}$ be the subspace of V consisting of the vectors that "transform under $\sigma(H)$ like the representation j' ". In case (5.11), the operator $\gamma^3 - \frac{i}{m}\gamma$ then maps $V^{j'}$ onto itself. The "masses" are then the values of m for which $\gamma^3 - \frac{i}{m}\gamma$ is not one-one, for some j'. j' is related to the "spin" - exactly how we shall investigate in Volume III. (5.12) is similar: The "spins" occurring in this "mass zero" case are the j' for which $\gamma_0 - \gamma_1 - i\gamma$ is not one-one.

The next step in this direction would be to investigate the relation between the numbers m, j, j' and the "mass" and "spin" as interpreted in terms of the Casimir operators of the universal enveloping algebra of the Poincaré Lie algebra. To carry this out at this moment - at least in a satisfactory way - would require a considerable detour to develop material from Lie algebra theory and the differential geometry of invariant differential operators on homogeneous vector bundles. We will leave this to a later point.

6. A GENERAL THEOREM ABOUT FINITE COMPONENT DIFFERENTIAL OPERATORS

We now want to prove some general results about the structure of the vector space V in case it is finite dimensional.

L will now denote the simply connected Lorentz group. (L is isomorphic to SL(2, C), of course, but to emphasize this fact notationally will only lead to confusion, since we want to treat L as a *real* Lie group). Let $\underset{\sim}{L}$ denote its Lie algebra, and let $\underset{\sim}{L}_c$ be its complexification, i.e. $\underset{\sim}{L}_c$ is the tensor product $\underset{\sim}{L} \otimes C$ (or, more naively just the linear combination of the generators of $\underset{\sim}{L}$ with complex coefficients.) Let L_c be the simply connected Lie group whose Lie algebra is $\underset{\sim}{L}_c$. (By general Lie theory, L_c exists and is unique, up to isomorphism).

<u>Exercise</u>. Show that L_c is isomorphic to SL(2, C) × SL(2, C). Make precise the complex-analytic manifold structure on SL(2, C) × SL(2, C) for which this isomorphism is an isomorphism in the sense of complex-analytic Lie groups.

Let $\sigma(L)$ be a representation of L by complex-linear operators on a complex vector space V, such that four operators γ^μ: $V \to V$ exist which transform like a 4-vector, i.e. like the $(\frac{1}{2}, \frac{1}{2})$ - representation, under $\sigma(L)$.

Let $\sigma(L)$ denote the corresponding representation of the Lie algebra $\underset{\sim}{L}$ by operators on V. Let (J_i, K_i) denote the usual basis of $\underset{\sim}{L}$. Then, the explicit conditions that (γ^μ) transform like $(\frac{1}{2}, \frac{1}{2})$ are the following:

$$
\begin{aligned}
&[\sigma(J_i), \gamma^o] = 0 \\
&[\sigma(J_i), \gamma^j] = \varepsilon_{ijk}\gamma^k \\
&[\sigma(K_i), \gamma^o] = \gamma^i \\
&[\sigma(K_j), \gamma^i] = \delta_{ij}\gamma^o
\end{aligned}
\qquad (6.1)
$$

σ can be extended to a representation σ_c of $\underset{\sim}{L}_c$ by the complex-linear condition:

$$\sigma_c(X + iY) = \sigma(X) + i\sigma(Y) \qquad (6.2)$$

for $X, Y \in \underset{\sim}{L}$.

By general Lie principles, σ_c arises as the infinitesimal generator of a representation - denoted also by σ_c - of L_c by operators on V.

Define operators δ^μ on V as follows:

$$\begin{aligned} \delta^o &= \gamma^o \\ \delta^i &= i\gamma^i \quad \text{for} \quad 1 \leq i \leq 3 \end{aligned} \tag{6.3}$$

Then, we see that:

$$\sigma_c(\ell_c)\delta^\mu\sigma_c(\ell_c{}^{-1}) = M_{\nu\mu}(\ell_c)$$

for $\ell_c \in L_c$ where $(M_{\mu\nu}(\ell_c))$ is an orthogonal 4×4 complex matrix,

i.e. an element of SO(4, C). Further, this assignment $\ell_c \to (M_{\mu\nu}(\ell_c))$ defines a homomorphism of $L_c \to$ SO(4, C).

LEMMA 6.1. This homomorphism $L_c \to$ SO(4, C) is *onto* SO(4, C).

Proof. From general Lie principles, the image of L_c in SO(4, C) is a connected subgroup of SO(4, C) of the same dimension as SO(4, C). From general principles again, it must be equal to SO(4, C) since SO(4, C) itself is connected. (See following exercise).

<u>Exercise</u>. Show that SO(4, C) is connected.

<u>Exercise</u>. In terms of the identification of SL(2, C) × SL(2, C) with L_c, determine this homomorphism into SO(4, C). Show explicitly that it is onto.

Consider now the following element of SO(4, C):

$$\begin{pmatrix} -1 & & & 0 \\ & -1 & & \\ & & -1 & \\ 0 & & & -1 \end{pmatrix} \tag{6.4}$$

i.e. the element representing "total reflection" in R^4 or C^4. (It is the product PT, where P is the "parity" and T the "time reversal" operator). By Lemma 6.1, there is an element $\ell_0 \in L_c$ which goes into the element (6.4) of SO(4, C). Explicitly, this means that:

$$\sigma(\ell_0)\delta^\mu\sigma(\ell_0^{-1}) = -\delta^\mu.$$

hence:

$$\beta\gamma^{\mu} = - \gamma^{\mu}\beta \tag{6.5}$$

where β is the linear transformation: $V \to V$ defined as follows:

$$\beta = \sigma(\ell_0) \tag{6.6}$$

<u>Exercise</u>. Any $\ell_0 \in L_c$ satisfying (6.6) satisfies:

$$\ell_0{}^2 = \text{identity} \tag{6.7}$$

$$\ell_0 \ell = \ell \ell_0 \quad \text{for all} \quad \ell \in L_c \tag{6.8}$$

Now we are prepared to analyze the conditions forced on V by the existence of the γ^{μ}. By (6.7), we have:

$$\beta^2 = \sigma(\ell_0)^2 = 1$$

Define subspace V_+, V_- of V, as follows:

$$\begin{aligned} V_+ &= \{v \in V: \ \beta(v) = v\} \\ V_- &= \{v \in V: \ \beta(v) = - v\} \end{aligned} \tag{6.9}$$

From (6.8), we have:

$$\sigma(L)(V_+) \subset V_+ \qquad (6.10)$$
$$\sigma(L)(V_-) \subset V_-$$

From (6.5), we have:

$$\gamma^\mu(V_+) \subset V_- \qquad (6.11)$$
$$\gamma^\mu(V_-) \subset V_+$$

Now, we have:

$$\sigma(\ell)\gamma = \gamma\sigma(\ell)$$

for all $\ell \in L$.

Hence, this relation holds over the complexes as well, i.e.

$$\sigma(\ell_c)\gamma = \gamma\sigma(\ell_c)$$

for $\ell_c \in L_c$

In particular:

$$\beta\gamma = \gamma\beta.$$

Hence:

$$\begin{aligned} \gamma(V_+) &\subset V_+ \\ \gamma(V_-) &\subset V_- \end{aligned} \tag{6.12}$$

Remarks. (6.11) immediately shows that the "four-vector" (γ^μ) cannot exist if the operator β-exists (as it does in the finite dimensional case) and if $\sigma(L)$ is irreducible. However, one should note that there are two irreducible representations, the "Majorana representations", of L which do admit such four-vectors. (The resulting invariant differential equations are the *Majorana equations*, see Ruhl [1] and Todorov [1]). These results indicate already that in the Majorana case $\sigma(L)$ cannot be extended to L_c, i.e. the "PCT-theorem" breaks down. (See Todorov [1]; Notice also that the fact that $\beta = \sigma(\ell_0)$, for $\ell_0 \in L_c$, is the key algebraic fact in the proof of the "PCT-theorem". See Streater-Wightman and Jost [1, 1]).

CHAPTER VIII

SOME GENERAL PROPERTIES OF POINCARÉ INVARIANT DIFFERENTIAL OPERATORS

One of the most interesting areas of application of modern mathematical ideas is the study of the mathematical patterns that seem to underly the spectrum of elementary particles and resonances. Here, the mathematical problems are not defined as sharply as they are in quantum field theory itself. What seems to be needed is some loose combination of quantum field theory ideas with the symmetry techniques that have been relatively successful in the isospin (SU(2)) and unitary spin (SU(3)) classification schemes for the elementary particles.

One such combination is the theory of "current algebras", an area that we hope to get into in more detail in later volumes of this work. Another area is that of the study of Poincaré invariant linear differential operators. Although this field does not seem to be as rich in physical results and ideas as "current algebras", it does give rise to an assortment of precise and interesting mathematical problems. In this chapter we begin to deal with some of these problems.

1. IRREDUCIBLE DIFFERENTIAL OPERATORS

Let L denote the simply connected Lorentz group SL(2, C), V denotes a complex vector space, and $\sigma(L)$ denotes a representation of L by operators on V. (γ^μ, γ) are operators on V, $0 \leq \mu, \nu \leq 3$, with the (γ^μ) and (γ) transforming like a "vector" and a "scalar" under $\sigma(L)$, i.e. via the $\left(\frac{1}{2}, \frac{1}{2}\right)$ and (0, 0) representation.

Then, $\Gamma(E)$ can be realized as the space of maps $\Psi\colon\ R^4 \to V$, with $D\colon\ \Gamma(E) \to \Gamma(E)$ defined by:

$$D\Psi = (\gamma^\mu \partial_\mu + \gamma)\Psi. \tag{1.1}$$

Let $\underset{\sim}{L}$ denote the Lie algebra of L, and let $\sigma(\underset{\sim}{L})$ denote the Lie algebra of operators on V obtained by passing from $\sigma(L)$ to the representation of the Lie algebra.

DEFINITION. The differential operator (1.1) is said to be *irreducible* if the Lie algebra generated by $\sigma(\underset{\sim}{L})$ the γ^μ, and γ is irreducible.

We will now present a few mathematical remarks that will be useful in the sequal.

DEFINITION. Let $\underset{\sim}{G}$ be a set of linear operators on a vector space V. The action of $\underset{\sim}{G}$ on V is *irreducible* if there is no subspace V' of V (other than V' = (0) or V' = V, of course) which is mapped into itself by all the operators in $\underset{\sim}{G}$.

If V is infinite dimensional, this definition is not really useful. It usually must be supplemented by certain "topological" assumptions concerning $\underset{\sim}{G}$ and V. However, in this chapter we will mainly be interested in the case where V is a finite dimensional vector space. In this case, the following basic structure theorem (due to E. Cartan) holds. We will not prove it here, but will refer to Jacobson [1]. A proof of Theorem 1.1 will be given in Volume II.

<u>THEOREM 1.1</u>. Let $\underset{\sim}{G}$ be a Lie algebra (under commutator) of linear transformations on a finite dimensional (complex) vector space V. Let $\underset{\sim}{C}$ be the center of $\underset{\sim}{G}$, i.e. the set of all $X \varepsilon \underset{\sim}{G}$ such that:

$$[X, \underset{\sim}{G}] = 0.$$

Then, there is an ideal $\underset{\sim}{S}$ of $\underset{\sim}{G}$ which is semisimple, such that:

$$\underset{\sim}{G} = \underset{\sim}{S} \oplus \underset{\sim}{C}, \tag{1.2}$$

i.e. $\underset{\sim}{G}$ is the direct sum of the semisimple Lie algebra $\underset{\sim}{S}$ and the abelian algebra $\underset{\sim}{C}$.

Further the dimension of $\underset{\sim}{C}$ is zero or one. If it is one, then $\underset{\sim}{C}$ consists of all multiples of the identity operator. Of course, in either case, $\underset{\sim}{S}$ acts irreducibly on V.

As examples where dim $\underset{\sim}{C} = 1$, we can cite the case where $\underset{\sim}{G}$ is the Lie algebra of GL(n, C) or U(n), acting on $C^n = V$. Then, $\underset{\sim}{S}$ is the Lie algebra of SL(n, C) or SU(n).

Now, let us return to the case of an irreducible first order Lorentz-invariant differential operator. Let $\underset{\sim}{H}$ be the real[1] Lie algebra of operators on V generated by $\sigma(\underset{\sim}{L})$ and the γ^μ.

Exercise: (Proved using Theorem 1.1.)

$\underset{\sim}{H}$ is a semisimple Lie algebra.

Let $\underset{\sim}{H}_c$ be the Lie algebra generated by the $\sigma(\underset{\sim}{L})$, γ^μ, and their commutators, with complex numbers as coefficients. Let (J_i, K_j), $1 \leq i, j \leq 3$, be the basis of L constructed before, with:

$$
\begin{aligned}
[J_i, J_j] &= \varepsilon_{ijk} J_k \\
[J_i, K_j] &= \varepsilon_{ijk} K_k \qquad (1.3) \\
[K_i, K_j] &= - \varepsilon_{ijk} J_k .
\end{aligned}
$$

Set:

[1]By this we mean that $\underset{\sim}{H}$ consists of the linear combinations of the $\sigma(\tilde{L})$, γ^μ and their commutators with real coefficients.

$$K_j' = iK_j$$

$$\gamma'^0 = \gamma^0 \qquad (1.4)$$

$$\gamma'^j = i\gamma^j.$$

Then, (J_i, K_j') generate a real Lie algebra, denote by: $\underset{\sim}{L}'$: (The real "compact form" of $\underset{\sim}{L}$, isomorphic to the Lie algebra of SO(4, R) and SU(2) × SU(2). Then,

$$[J_i, \gamma'^0] = 0$$

$$[J_i, \gamma'^j] = \varepsilon_{ijk}\gamma'^k$$

$$[K_i', \gamma'^0] = \gamma'^j \qquad (1.5)$$

$$[K_i', \gamma'^j] = -\delta_{ij}\gamma'^0.$$

Let $\underset{\sim}{H}$ be the real Lie algebra generated by the J_i, K_i', γ^μ. It is then clear that $\underset{\sim}{H}$ is a semisimple, real Lie algebra containing the compact Lie algebra $\underset{\sim}{L}'$, and having the property that Ad $\underset{\sim}{L}'$ acting in $\underset{\sim}{H}'$ has - in the decomposition into irreducible representations - a "vector" representation, i.e. a $\left(\frac{1}{2}, \frac{1}{2}\right)$.

This gives us a "structure theorem" for all irreducible Poincaré invariant differential operators. They are obtained by choosing real semi-simple Lie algebras $\underset{\sim}{H}'$ containing $\underset{\sim}{L}'$, having the property that Ad $\underset{\sim}{L}'$ acting in $\underset{\sim}{H}$ transforms a subspace like a 4-vector, and then choosing an irreducible representation of $\underset{\sim}{H}'$. In principle then, all such operators are classified, although admittedly carrying out this classification would involve one deeply in the technicalities of Lie algebra theory; certainly it does not yet exist in the literature.

However, the main advantage to us in describing these operators in this way is that it suggests a general method for constructing interesting examples of this type of operator. Indeed, there is nothing preventing us from using this method to construct "infinite component" differential operators - select $\underset{\sim}{H}$, and then choose any representation, finite or infinite dimensional, to define V. This method has been extensively used by Nambu [1] to construct operators with interesting properties.

Example: *The Dirac equation.*

We will now show that the Dirac equation corresponds - as explained in the last section - to the imbedding of SL(2, C) in the Lie group SU(2, 2).

First, we will construct SU(2, 2) in a certain way. Let V' be a 2-dimensional, complex-Hilbert space, with the Hilbert inner product denoted by $<v_1'/v_2'>$, for $v_1', v_2' \in V'$.

Set:

$$V = V' \oplus V'.$$

Let us define an Hermitian - symmetric inner product[1] on V as follows.

$$<v_1' + v_2'/v_3' + v_4'> = <v_1'/v_3'> - <v_2'/v_4'>. \tag{1.6}$$

[1]If V is a complex-vector space, a *Hermitian symmetric* inner product is defined as follows: It is a real-bilinear map $(v_1, v_2) \to <v_1/v_2>$ such that: $<v_1/v_2> = <v_2/v_1>^*$;

$$<cv_1/v_2> = c^*<v_1/v_2>; \quad <v_1/cv_2> = c<v_1/v_2>$$

for $c \in C$; $v_1, v_2 \in V$. * denotes complex-conjugate.

Let us recall some general mathematical material about Hermitian symmetric forms. Let V be a complex vector space, and let $(v_1, v_2) \to \langle v_1/v_2\rangle$ be a Hermitian symmetric form on V. Let us suppose that it is *nondegenerate*, i.e. $\langle v/V\rangle = 0$ implies $v = 0$.

<u>Exercise</u>: Show that there exist two linear subspaces V_1, V_2 of V such that:

$$V = V_1 \oplus V_2.$$

$$\langle V_1/V_2\rangle = 0$$

$$\langle v_1/v_1\rangle > 0 \quad \text{for all} \quad v_1 \in V_1$$

$$\langle v_2/v_2\rangle < 0 \quad \text{for all} \quad v_2 \in V_2.$$

Let us now suppose that (V_1, V_2) and (V_1', V_2') are two pairs of such subspaces.

<u>Exercise</u>: Show that there exists an isomorphism $A\colon V \to V$ of V onto itself such that:

$$\langle Av_1/Av_2\rangle = \langle v_1/v_2\rangle$$

for $v_1, v_2 \in V$.

$$A(V_1) = V_1'; \; A(V_2) = V_2'.$$

This exercise then says that this splitting of V is unique, up to isomorphism.

<u>Exercise</u>: Suppose: $\dim V_1 = p$:, : $\dim V_2 = q$: Show that there exist bases (v_i), $1 \leq i, j \leq p$; (v_a), $1 \leq a, b \leq q$ of V_1 and V_2 such that:

$$\langle v_i/v_j \rangle = \delta_{ij}$$

$$\langle v_a/v_b \rangle = -\delta_{ab}.$$

Thus, if $v = c_i v_i + c_a v_a$, with $c_i, c_a \in C$, then:

$$\langle v/v \rangle = c_i c_i^* - c_a c_a^*.$$

Thus, the "canonical form" for the form $\langle 1 \rangle$ has p plus and q minus signs. These numbers (p, q) determine the form up to isomorphism. The group of isomorphism of two such forms having the same (p, q) are then isomorphic; in Lie group theory one uses the notation: U(p, q): for this group. The form (1.6) has in its canonical 2 plus and 2 minus signs; therefore, we call its group of automorphisms U(2, 2), and the subgroup of determinant one automorphisms SU(2, 2).

If A: $V \to V$ is a complex-linear transformation, let A^* denote the Hermitian adjoint of A with respect to the form (1.6). Then, the Lie algebra of SU(2, 2) consists of $A \in L(V)$ [1] such that: $A^* = -A$ and trace (A) = 0:

Define two subspaces, V_+ and V_-, of V, in the following way:

$$V_+ = \{v' \oplus v' : \quad v' \in V'\}$$
$$V_- = \{v' \oplus -v' : \quad v' \in V\}. \tag{1.7}$$

Then, V_+ and V_- are *isotropic subspaces* of V, i.e.

$$\langle v_+/v_+\rangle = 0 = \langle v_-/v_-\rangle$$
$$\text{for} \quad v_+, v_- \in V_+, V_-. \tag{1.8}$$

Now, let L = SL(2, C) = simply connected covering group of the Lorentz group. Regard L as

[1] L(V) denotes the Lie algebra (under commutator) of the linear transformations: $V \to V$.

the group complex linear transformations $\ell: V' \to V'$ of determinant one. Let ℓ^* denote the adjoint operator to ℓ, with respect to the given Hilbert space structure on V'.

Let ℓ define linear transformations: $V_+ \to V_+$, $V_- \to V_-$, in the following way:

$$\begin{aligned} \ell(v' \oplus v') &= \ell(v') \oplus \ell(v') \\ \ell(v' \oplus -v') &= \ell^{*-1}(v') \oplus -\ell^{*-1}(v'). \end{aligned} \tag{1.9}$$

Exercise: Show that ℓ' defined by (1.9) as acting on $V = V_+ \oplus V_-$ is an element of SU(2, 2), i.e. ℓ' preserves the inner product (1.6).

Exercise: Calculate a formula similar to (1.9) showing how the Lie algebra of SL(2, C) is imbedded as a subalgebra of the Lie algebra of SU(2, 2).

Exercise: With $\underset{\sim}{L}$ imbedded - as explained in the exercise - as a subalgebra of $\underset{\sim}{H}$, where H = SU(2, 2), show that there is a subspace of $\underset{\sim}{H}$, which transforms like the $\left(\frac{1}{2}, \frac{1}{2}\right)$ - representation under $\underset{\sim}{L}$.

Hint: If $\gamma^\mu \in \underset{\sim}{H}$ generate the subspace in question, show that:

$$\gamma^{\mu}(V_{+}) \subset V_{-}; \quad \gamma^{\mu}(V_{-}) \subset V_{+}.$$

Exercise: Show that those constructions lead to the physicists' version of the Dirac equation, with γ^{μ} identified with the 4×4 Dirac matrices. This can be done by choosing appropriate bases for the vector spaces constructed above.

Exercise: If $D = \gamma^{\mu}\partial_{\mu} + im$ is the Dirac differential operator, show that the solutions of: $D\psi = 0$: are solutions of the Klein-Gordon differential equation:

$$g_{\mu\nu}\partial_{\mu}\partial_{\nu} + m^2 = 0.$$

2. GENERAL REMARKS ABOUT MASS SPECTRUM OF FINITE COMPONENT EQUATIONS

Let us now work with the following notations: $\underset{\sim}{L}$ denotes the Lie algebra of the Lorentz group; $\sigma(\underset{\sim}{L})$ is a representation of $\underset{\sim}{L}$ by operators on a finite dimensional vector space, V; (γ^μ) are a set of operators on V that transform under $\sigma(\underset{\sim}{L})$ like a "vector", i.e. like the $\left(\frac{1}{2}, \frac{1}{2}\right)$ -representation.

Construct the following Poincaré-invariant differential operator:

$$D = \gamma^\mu \partial_\mu - im, \tag{2.1}$$

where m is a non-zero real scalar. Our goal is to study the mass-spin spectrum of this equation.

Let (J_i, K_j) be the usual basis of $\underset{\sim}{L}$. Set:

$$K_j' = iK_j$$

$$\gamma'^\mu = \begin{Bmatrix} \gamma^0 & \text{if} \quad \mu = 0 \\ \\ i\gamma^j & \text{if} \quad \mu = j \end{Bmatrix}. \tag{2.2}$$

Let $\underset{\sim}{L}'$ be the Lie algebra generated by the (J_i, K_j'). As we have seen, it is the Lie algebra of the simply connected group: SU(2) × SU(2); that we denote by: L': Since L' is simply connected, σ defines a representation - that we again denote by σ(L') - of L' by operators on V. Since

L' is compact, there is a Hilbert-space inner product[1] denoted by $<v_1/v_2>$, on V, with respect to which $\sigma(L')$ consists of unitary transformations.

Let us suppose that the following condition is satisfied:

γ^0 is a Hermitian operator with respect to the form $< \mid >$, i.e. all the eigenvalues of γ^0 are real numbers. (2.3)

With those assumptions, we are prepared to make some general statements about the mass-spin spectrum.

Let $\underset{\sim}{L}_c = \underset{\sim}{L} + i\underset{\sim}{L} = \underset{\sim}{L}' + i\underset{\sim}{L}'$ be the complex Lie algebra generated by $\underset{\sim}{L}$ and $\underset{\sim}{L}'$. Let L_c be the corresponding simply connected Lie group. Again, from general Lie theory we know that σ can be extended to a representation of L_c. Then, for $\ell_c \in L_c$,

$$\sigma(\ell_c)(\gamma^\mu) = M_{\nu\mu}(\ell_c)\gamma^\nu, \tag{2.4}$$

---

[1]That is, a positive definite, Hermitian symmetric bilinear form: $V \times V \to C$.

where $\ell_c \to (M_{\mu\nu}(\ell_c))$ is the $\left(\frac{1}{2}, \frac{1}{2}\right)$ - matrix representation of L_c. Now, $(M_{\mu\nu}(\ell_c))$ is an element of SO(4, C). (2.3) then proves the following result:

<u>THEOREM 2.1</u>. For $p = (p_\mu) \in C^4$, set:

$$\gamma(p) = p_\mu \gamma^\mu.$$

If $p, p' \in C^4$, with

$$p^2 = g_{\mu\nu} p_\mu p_\nu = p'^2,$$

and with $p, p' \neq 0$, then

$$\gamma(p) \text{ and } \gamma(p')$$

have the same eigenvalues. In fact, there is an $\ell_c \in L_c$ such that:

$$\sigma(\ell_c)\gamma(p)\sigma(\ell_c)^{-1} = \gamma(p').$$

<u>THEOREM 2.2</u>. If (2.3) holds, then the operators γ^j are skew-Hermitian, hence all of their eigenvalues are pure imaginary.

The operators $\gamma^j - \gamma^0$ are nilpotent, i.e. all their eigenvalues are zero.

Proof. By the construction of the Hilbert space on V, the operators $\sigma(K_j{}')$ are skew-Hermitian; hence the operators $\sigma(K_j)$ are Hermitian, since they are related by:

$$\sigma(K_j) = i\sigma(K_j{}').$$

Now,

$$\gamma^j = [\sigma(K_j), \gamma^0].$$

Recall that the commutator of two Hermitian operators is skew-Hermitian, hence γ^1 is skew-Hermitian.

To deal with the second part of the theorem, note that:

$$[\sigma(K^j), \gamma^j - \gamma^0] = -(\gamma^j - \gamma^0).$$

It is now an exercise in Lie algebra theory (left to the reader) to show that this relation implies that $\gamma^j - \gamma^0$ is a nilpotent transformation.

Now, let us study the mass-spin spectrum of the operator (2.1). Let $c_1, \ldots, c_n$ be the eigenvalues of γ^0, with $V_1, \ldots, V_n$ the corresponding eigenvector spaces. Since γ^0 is a Hermitian operator, these eigenspaces are mutually orthogonal, and V is a direct sum of them. Let K be the subgroup of L generated by the subalgebra (J_i). Then,

$$\sigma(K)(V_\alpha) \subset V_\alpha, \quad \alpha = 1, \ldots, n. \tag{2.5}$$

Let $\Gamma(E)$ denote the space of map Ψ: $R^4 \to V$, $\Gamma(E, D)$ those which are annihilated by the operator D given by (2.1). Consider a $\Psi \in \Gamma(E)$ of the following form:

$$\Psi(x) = e^{ip \cdot x} v, \tag{2.6}$$

with $p \in C^4$, $v \in V$.

Then, $\Psi \in \Gamma(E, D)$ if and only if:

$$\gamma(p)(v) = mv. \tag{2.7}$$

Let $V(p)$ denote the space of vectors v satisfying (2.7), and let M be the set of $p \in C^4$ such that

$V(p) \neq (0)$.

Let SO(1, 3; C) be the group of linear transformstions: $C^4 \to C^4$ that preserve the form: $p^2 = q_{\mu\nu}p_\mu p_\nu$:, and have determinant +1. Then, M is invariant under the action of SO(1, 3; C) on C^4. Let N be an orbit of this group acting on M. Let $x = p^2$, where p is a point on N.

Case 1. $x \neq 0$.

Then, the point $(\sqrt{x}, 0, 0, 0) = p^0$ is in N. Hence, $V(p^0) \neq (0)$. Hence, x is one of the following numbers:

$$x = \frac{m^2}{m_1^2}, \ldots, \frac{m^2}{m_n^2}. \tag{2.8}$$

(In particular, note that x, the "mass" defined by the Poincaré group, is positive. No "negative-mass" particles are present.)

Case 2. $x = 0$.

This case cannot occur. For, then the point $(-1, 1, 0, 0) = p^0$ would be in N. If $V(p^0) \neq (0)$, $\gamma^1 - \gamma^0$ would have non-zero eigenvalues, contradicting Theorem 2.2.

Let us then return to the study of case 1, considering only the points $M \cap R^4$. (See the next section for a discussion of this point.) The "spins" are then identified with the irreducible representations of the isotropy subgroup of L at the point p.

Case 1. $p_0 > 0$.

Then, we can refer back (via the action of the *real* group $SO^+(1, 3)$) to the point $(\sqrt{x}, 0, 0, 0)$. Suppose, for example, that

$$x = \frac{m^2}{m_1^2} .$$

Then, the "spins" are the irreducible representations of K acting on the vectors annihilated by the following operator:

$$\gamma^0 - |m_1| .$$

In particular, if $m_1 > 0$, then it is defined by the action of $\sigma(K)$ in V_1. If $m_1 < 0$, then they are defined by the action of $\sigma(K)$ in V_2, where m_2 is the eigenvalue such that: $m_2 = -m_1$.

Case 2. $p_0 < 0$.

The situation is reversed, but analogous. We refer via $SO^+(1, 3)$ back to the point $(-\sqrt{x}, 0, 0, 0)$.

To complete this discussion, we prove:

THEOREM 2.3. Suppose that $p = (p_0, 0, 0, 0)$, $p' = (-p_0, 0, 0, 0)$. Then, there is an element $\ell_0 \in L_c$ such that:

$$\gamma(p) = \sigma(\ell_0)\gamma(p')\sigma(\ell_0)^{-1} \tag{2.9}$$

i.e. γ^0 and $-\gamma^0$ have the same set of eigenvalues.

Proof. Again, this follows from two facts:

a) Since V is finite dimensional, σ-can be extended from L to L_c.

b) The "total reflection" or "PT" operation: $p \to -p$ is in $SO(1, 3; C)$, and this group is connected, hence this transformation can be accomplished - by a general Lie theory - by an element in L_c.

3. UNITARY REPRESENTATIONS OF THE POINCARÉ GROUP DEFINED BY DIFFERENTIAL OPERATORS

Let V be a complex vector space, let $\Gamma(E)$ denote the space of all maps Ψ: $R^4 \to V$, and let D: $\Gamma(E) \to \Gamma(E)$ be a differential operator. Let L be the simply connected Lorentz group, $\sigma(L)$ a representation of L by operators on V, $G = L \cdot T$ the Poincaré group, and with ρ the representation of G by operators on $\Gamma(E)$ defined by σ.

Suppose that D commutes with the action of $\rho(G)$. Then, $\rho(G)$ maps onto itself $\Gamma(E, D)$ the space of $\Psi \in \Gamma(E)$ such that:

$$D\Psi = 0$$

Our problem: When can $\Gamma(E, D)$ be given a Hermitian-symmetric inner product that is invariant under the action of $\rho(G)$?

Now, we are not interested in this from a purely "abstract" point of view. To be useful in quantum field theory, it seems to desirable that this inner product on $\Gamma(E, D)$ be defined by a "conserved current". In the language of differential operators, this can be explained as follows:

Let $V_c(R^4)$ denote the vector fields on R^4 with complex functions as coefficients.

Let $\omega\colon\ \Gamma(E) \times \Gamma(E) \to V_c(R^4)$ be a real bilinear differential operator such that:

$$\begin{aligned} &\text{a)}\quad \omega(\Psi_1, \Psi_2) = \omega(\Psi_2, \Psi_1)^* \\ &\text{b)}\quad \omega(c\Psi_1, \Psi_2) = c^*\omega(\Psi_1, \Psi_2) \\ &\qquad\ \ \omega(\Psi_1, c\Psi_2) = c\omega(\Psi_1, \Psi_2) \end{aligned} \tag{3.1}$$

for $\Psi_1, \Psi_2 \in \Gamma(E)$, $c \in C$.

Let $\omega_\mu(\Psi_1, \Psi_2)$ be the functions of $x \in R^4$ describing the components of ω with respect to the basis ∂_μ $(= \frac{\partial}{\partial x_\mu})$ of $V(R^4)$.

$$\omega(\Psi_1, \Psi_2) = \omega_\mu(\Psi_1, \Psi_2)\partial_\mu$$

We are now looking for an ω satisfying:

$$\partial_\mu(\omega_\mu(\Psi_1, \Psi_2)) = 0 \tag{3.2}$$

for $\Psi_1, \Psi_2 \in \Gamma(E, D)$.

Using such a ω, define an inner product $< / >$ on $\Gamma(E, D)$ as follows:

$$<\Psi_1/\Psi_2> = \int \omega_0(\Psi_1, \Psi_2)(0, \vec{x})d\vec{x}, \tag{3.3}$$

where $\vec{x}$ denotes (Ψ_1, Ψ_2, Ψ_3) and $d\vec{x}$ is the Euclidean volume element form $dx_1 \wedge dx_2 \wedge dx_3$.

Of course, (3.3) singles out the "zero-time" hypersurface: $x_0 = 0$: of R^4. The "conserved" condition, (3.2), guarantees however - as we have already seen - that the integration on the right hand side of (3.3) can just as well - by Stokes' formula - be taken over any hypersurface in R^4 which - together with the hypersurface: $x_0 = 0$: is the boundary of a region in R^4 in which the intergrand is C^∞. In addition, one must specify growth conditions at infinity - "compact support" would be the simplest, - for the Ψ_1, Ψ_2 which guarantee that the integrands converge and that the integration by parts that are implicit in Stokes' formula are justified. In fact, we will usually only be working "formally"[1] and ignoring possible analytical subtilties.

We now have two tasks: Investigate in more detail the conditions which assure that $\rho(G)$ leaves invariant the inner product defined by (3.3). Once this is done, we will further investigate the relation to the induced-representations analysis of $\Gamma(E, D)$ performed in previous work, and inquire under what conditions the inner product $\langle \mid \rangle$ is positive definite.

For simplicity, let us work with the simplest general form for D; namely let us suppose that it is a first-order differential operator, of the form.

$$D = \gamma^{\mu}\partial_{\mu} + \gamma \qquad (3.4)$$

where (γ^{μ}, γ) are operators: $V \to V$ that transform under $\sigma(L)$ like a "vector" and a "scalar".

---

[1]As this term is used in the mathematical literature - and it differs from its use in physics - it usually means that possibly-divergent intergrals and tricky integration - by parts are ignored. The point is that we want to understand the algebraic features of the formalism, leaving until later the more delicate analytical aspects.

Let us look for an ω satisfying (3.1) that is a zeroth order differential operator. (In general, it is clear from (3.2) that it is reasonable to expect ω to be of one order lower than D, since the operator "div" is a first order operator.) To put this in the most convenient form, suppose that V has a Hermitian symmetric bilinear form, denoted by $(v_1, v_2) \to (v_1, v_2)$, with $\omega_\mu(\Psi_1, \Psi_2)$ defined as follows:

$$\omega_\mu(\Psi_1, \Psi_2)(x) = (\beta_\mu \Psi_1(x), \Psi_2(x)) \tag{3.5}$$

where (β_μ) are operators: $V \to V$ that are Hermitian with respect to the form (v_1, v_2), i.e.

$$(\beta_\mu v_1, v_2) = (v_1, \beta_\mu v_2) \tag{3.6}$$

for $v_1, v_2 \in V$.

Now we can satisfy (3.2) *algebraically* by means of a relation of the following form:

$$\partial_\mu(\beta_\mu \Psi_1, \Psi_2) = (\alpha_1 \Psi_1, D\Psi_2) + (D\Psi_1, \alpha_2 \Psi_2), \tag{3.7}$$

where α_1, α_2 are linear maps: $V \to V$. Let us work out both sides of (3.4), and compare coefficients: The left hand side is:

$$(\beta_\mu \partial_\mu \Psi_1, \Psi_2) + (\beta_\mu \Psi_1, \partial_\mu \Psi_2)$$

The right hand side is:

$$\begin{aligned}&(\alpha_1 \Psi_1, \gamma^\mu \partial_\mu \Psi_2) + (\alpha_1 \Psi_1, \gamma \Psi_2) + (\gamma^\mu \partial_\mu \Psi_1, \alpha_2 \Psi_2)\\ &+ (\gamma \Psi_1, \alpha_2 \Psi_2) = (\gamma^{\mu *} \alpha_1 \Psi_1, \partial_\mu \Psi_2)\\ &+ (\gamma^* \alpha_1 \Psi_1, \Psi_2) + (\partial_\mu \Psi_1, \gamma^{\mu *} \alpha_2 \Psi_2)\\ &+ (\alpha_2{}^* \gamma \Psi_1, \Psi_2)\end{aligned}$$

(Here * denotes the adjoint of an operator on V with respect to the form (,)). Comparing, we have:

$$\gamma^* \alpha_1 = \alpha_2{}^* \gamma \tag{3.8}$$

$$\gamma^{\mu *} \alpha_1 = \beta_\mu \tag{3.9}$$

$$\beta_\mu{}^* = \gamma^{\mu *} \alpha_2 \tag{3.10}$$

Conversely, these conditions are reversible,

leading to formula (3.7).

Let us reduce formulas (3.8) - (3.10) further. Apply * to both sides of (3.10), and equate to (3.9):

$$\beta_\mu = \alpha_2{}^*\gamma^\mu = \gamma^{\mu *}\alpha_1 \tag{3.11}$$

Thus, we have proved:

<u>THEOREM 3.1</u>. Suppose α_1, α_2 are operators: $V \to V$ satisfying the following conditions:

$$\begin{aligned} \gamma^*\alpha_1 &= \alpha_2{}^*\gamma \\ \gamma^{\mu *}\alpha_1 &= \alpha_2{}^*\gamma^\mu \end{aligned} \tag{3.12}$$

Then, if β_μ is defined by the following formula:

$$\beta_\mu = \alpha_2{}^*\gamma^\mu. \tag{3.13}$$

$\omega_\mu(\Psi_1, \Psi_2)$ defined by (3.5) also satisfies (3.7). If in addition:

$$\alpha_2{}^*\gamma = \gamma^*\alpha_2, \tag{3.14}$$

then β is Hermitian, hence (3.3) defines a Hermitian-symmetric form on $\gamma(E, D)$.

Let us work out more explicitly formulas for α_1, α_2 in various cases:

<u>Case 1</u>. γ is skew-Hermitian; γ^o is Hermitian; γ^i is skew-Hermitian.

For example, this is the situation for the Dirac equation.

Explicitly, we have:

$$\gamma^* = -\gamma; \; \gamma^* = -\gamma^o; \; \gamma^{i*} = -\gamma^i$$

Then, (3.12), (3.14) take the following form:

$$\begin{aligned} \gamma\alpha_1 &= -\alpha_2^*\gamma \\ \gamma^o\alpha_1 &= \alpha_2^*\gamma^o = \gamma^o\alpha_2 \\ -\gamma^j\alpha_1 &= \alpha_2^*\gamma^j \end{aligned} \tag{3.15}$$

<u>Case 2</u>. γ is skew-Hermitian, γ^μ are Hermitian: Then,

$$\gamma\alpha_1 = -\alpha_2^*\gamma$$

$$\gamma^\mu\alpha_1 = \alpha_2^*\gamma^\mu \tag{3.16}$$

$$\alpha_2\gamma^o = \gamma^o\alpha_2$$

Now, we can ask for the conditions that the inner product (3.3) be invariant under the actions of the Poincaré group on $\Gamma(E, D)$. The answer goes as follows:

<u>THEOREM 3.2</u>. Suppose the operators (β^μ) transform as a four-vector under the action of the Lorentz group on V. Also, suppose (3.2) is satisfied. Then, the action of the Poincaré group on $\Gamma(E, D)$ preserves the inner product (3.3).

The direct proof of this is left as an exercise. Instead, of working further in these directions, we will detour in order to redevelop this material in stricter manifold-differential geometric language. This detour will benefit in two ways: First, it provides a means of generalizing much of this material (which is of great interest for mathematics, if not for physics) to more complicated differential-geometric situations. Second, it will give added geometric insight to even the simple situations that have so far been encountered in physically-interesting situations.

4. CONSERVED CURRENTS ASSOCIATED WITH GENERAL DIFFERENTIAL OPERATORS

As indicated in Section 3, our goal is now to develop some of the ideas developed there in a general context. Let $\pi: E \to M$ be a vector bundle over a manifold M. Let $F_c^r(M)$, $r = 0, 1, 2, \ldots$ denote the F(M)-module of complex-valued r - differential form on M. As usual, let $\Gamma(E)$ denote the F(M)-module of cross-sections of the vector bundle E. In this section, the fibers of E will be complex vector spaces.

For the purposes of this section, at least, a *current* - typically denoted by ω - will be a real-bilinear map $\omega: \Gamma(E) \times \Gamma(E) \to F_c^{m-1}(M)$ which is Hermitian-symmetric, i.e.

$$\begin{aligned} \omega(\Psi_1, \Psi_2) &= \omega(\Psi_2, \Psi_1)^* \\ \omega(c\Psi_1, \Psi_2) &= c^*\omega(\Psi_1, \Psi_2) = \omega(\Psi_1, c^*\Psi_2) \end{aligned} \tag{4.1}$$

for $\Psi_1, \Psi_2 \in \Gamma(E)$, $c \in C$. (m = dimension M).

Let N be on (oriented) submanifold of M of dimension m - 1. For $\Psi_1, \Psi_2 \in \Gamma(E)$, define:

$$\underset{\sim}{\omega}(\Psi_1, \Psi_2; N) = \int_N \omega(\Psi_1, \Psi_2) \tag{4.2}$$

(The right hand side is the integral over N of the (n - 1)-form resulting from restricting $\omega(\Psi_1, \Psi_2)$ to N). Thus, $\underset{\sim}{\omega}$ is a "function" of Ψ_1, Ψ_2, N. For $\Psi \in \Gamma(E)$, define:

$$\underset{\sim}{\omega}(\Psi; N) = \underset{\sim}{\omega}(\Psi, \Psi; N) \tag{4.3}$$

Thus, $\underset{\sim}{\omega}(\Psi; N)$ is a "function" assigning a real number to a $\Psi \in \Gamma(E)$ and a submanifold N of M. Physically, this real number may be thought of as the "charge" associated with the "field" Ψ on the submanifold M.

Let D: $\Gamma(E) \to \Gamma(E)$ be a linear differential operator. Let $\Gamma(E, D)$ denote the set of $\Psi \in \Gamma(E)$ such that: $D\Psi = 0$: Let $\Psi_1, \Psi_2 \to h(\Psi_1, \Psi_2)$ denote an F(M)-linear, Hermitian symmetric inner product, mapping: $\Gamma(E) \times \Gamma(E) \to F_c(M)$. Let α_1, α_2 be $F_c(M)$-linear maps: $\Gamma(E) \to \Gamma(E)$. Then, ω is a *conserved current* if α_1, α_2 and the inner product $h(\Psi_1, \Psi_2)$ can be chosen so that:

$$d\omega(\Psi_1, \Psi_2) = [h(\alpha_1\Psi_1, D\Psi_2) + h(D\Psi_1, \alpha_2\Psi_2)]dp \tag{4.4}$$

for $\Psi_1, \Psi_2 \in \Gamma(E)$

(dp denotes a fixed volume-element form for M).

In particular, (4.4) obviously implies that:

$$d\omega(\Psi_1, \Psi_2) = 0 \tag{4.5}$$

for Ψ_1 and $\Psi_2 \in \Gamma(E, D)$

In turn, (4.5) - together with Stokes' theorem - implies that the "charge" $\underset{\sim}{\omega}(\Psi; N)$, for given $\Psi \in \Gamma(E, D)$, takes the same value when N varies with a class of submanifolds of M which co-bound a region of M. (For example, if $M = R^4$, the submanifolds of the form $\{x \in R^4\colon\ x_0 = \text{constant}\}$, i.e. the "time-slice" submanifolds of R^4, obviously form such a class.)

Now, we must study the group - theoretic

properties of these objects. Let G be a Lie group. Suppose that G acts linearly on E and on the base space M, leaving invariant the differential operator D. Let $\underset{\sim}{G}$ be the Lie algebra of G. Thus $X \in \underset{\sim}{G}$ acts in two ways via the Lie derivative operation:

a) As a first order, linear differential operator

$$\Psi \to X(\Psi) \text{ on } \Gamma(E) \tag{4.6}$$

b) By Lie derivative $\omega \to X(\omega)$ on differential forms M.

DEFINITION. A current-type mapping $\omega\colon \ \Gamma(E) \times \Gamma(E) \to F_c^{m-1}(M)$ is said to be *G-invariant* if ω intertwines the natural action of G on $\Gamma(E)$ and $F_c^{m-1}(M)$.

Thus, at the Lie algebra level, G-invariance means that:

$$X(\omega(\Psi_1, \Psi_2)) = \omega(X(\Psi_1), \Psi_2) + \omega(\Psi_1, X(\Psi_2)) \tag{4.7}$$

for $\Psi_1, \Psi_2 \in \Gamma(E)$, $X \in \underset{\sim}{G}$.

Let us suppose now that (4.7) is satisfied. For $X \in \underset{\sim}{G}$, set:

$$\omega_X(\Psi_1, \Psi_2) = \grave{\omega}(X(\Psi_1), \Psi_2) \tag{4.8}$$

Then,

$$\omega_X(\Psi_1, \Psi_2) - \omega_X(\Psi_2, \Psi_1)^* = \omega(X(\Psi_1), \Psi_2)$$

$$- \omega(X(\Psi_2), \Psi_1)^* = \text{, using (4.7),}$$

$$\omega(X\Psi_1, \Psi_2)[X(\omega(\Psi_2, \Psi_1) - \omega(\Psi_2, X\Psi_1)]^*$$

$$= X(\omega(\Psi_1, \Psi_2)) \tag{4.9}$$

In particular, if N is a submanifold of M of dimension $(m - 1)$, set:

$$\underset{\sim}{\omega}(\Psi, X, N) = \int_N \omega_X(\Psi, \Psi) \tag{4.10}$$

<u>THEOREM 4.1</u>. If Stokes' theorem is applicable, i.e. if the data falls off sufficiently rapidly near the "boundary" of N, then $\underset{\sim}{\omega}(\Psi, X, N)$ is a real number for $\Psi \in \Gamma(E, D)$. Physically, this real number is the "charge" associated with the "current"

defined by X.

<u>Proof</u>. The basic identities relating Lie derivative, exterior derivative and inner product gives:

$$X(\omega(\Psi, \Psi)) = X \lrcorner\, d\omega(\Psi, \Psi) + d(X \lrcorner\, \omega(\Psi, \Psi)). \tag{4.11}$$

Since ω is a "conserved current" associated with differential operator D, and since $D\Psi = 0$, we have:

$$d\omega(\Psi, \Psi) = 0,$$

and

$$\int_N X(\omega(\Psi, \Psi)) = \int_N d(X \lrcorner\, \omega(\Psi, \Psi)) = 0,$$

if Stokes' theorem is applicable (4.11) and (4.9) then imply that $\underset{\sim}{\omega}(\Psi, X, N)$ is a real number.

Then, this assignment $\Psi \to \underset{\sim}{\omega}(\Psi, X, N)$ defines a real valued function on $\Gamma(E, D)$. In the language of quantum mechanics, this function is the "observable" associated with the "one parameter group of symmetries" generated by X. (See "Lie algebras and quantum mechanics" [4] for a general description of this relation).

5. CONSERVED CURRENTS ASSOCIATED WITH LAGRANGIANS

The machinery described in the previous sections takes a maximally elegant form in case the differential operator D is associated with a Lagrangian L. Suppose that E, E' are vector bundles over M. Suppose that the vector bundles E and E' over M have Hermitian-symmetric inner products on the fibers, denoted by:

$$\Psi_1, \Psi_2 \to h(\Psi_1, \Psi_2)$$

$$\Psi_1', \Psi_2' \to h'(\Psi_1', \Psi_2')$$

Suppose that D_1, D_2 are differential operators: $\Gamma(E) \to \Gamma(E')$, and L, a "Lagrangian", has the following form:

$$L(\Psi) = \int_M h'(D_1\Psi, D_2\Psi)dp \qquad (5.1)$$

Then, D, the Euler-Lagrange differential operator

associated with L, has the form:

$$D = D_1^* D_2 + D_2^* D_1 \tag{5.2}$$

Where D_1^*, D_2^* denote the adjoint operators defined by the Hermitian forms h and h', i.e.

$$\int_M h'(D_1\Psi_1, \Psi_2')dp = \int_M h(\Psi_1, D_1^*\Psi_2') \tag{5.3}$$

$$\text{for} \quad \Psi_1 \in \Gamma_0(E), \Psi_2 \in \Gamma_0(E')$$

Let us suppose that (5.3) arises from an identity of the following form:

$$[h'(D_1\Psi_1, \Psi_2') - h(\Psi_1, D_1^*\Psi_2')]dp$$

$$= d\theta_1(\Psi_1, \Psi_2') \tag{5.4}$$

$$\text{for} \quad \Psi_1, \in \Gamma(E), \Psi_2' \in \Gamma$$

where θ_1 is a bilinear differential operator: $\Gamma(E) \times \Gamma(E') \to F_c^{m-1}(M)$. Assume a similar relation for D_2:

$$[h'(D_2\Psi_1, \Psi_2') - h(\Psi_1, D_2^*\Psi_2')]dp$$
$$= d\theta_2(\Psi_1, \Psi_2') \tag{5.5}$$

Using (5.4) and (5.5), we have, for $\Psi_1, \Psi_2 \in \Gamma(E)$,

$$d\theta_1(\Psi_1, D_2\Psi_2) = [h'(D_1\Psi_1, D_2\Psi_2)$$

$$- h(\Psi_1, D_1^* D_2\Psi_2)]dp$$

$$d\theta_2(\Psi_1, D_1\Psi_2) = [h'(D_2\Psi_1, D_1\Psi_2)$$

$$- h(\Psi_1, D_2^*\Psi_2)]dp$$

Thus, we have:

$$d\theta_1(\Psi_1, D_2\Psi_2) + d\theta_2(\Psi_1, D\Psi_2)$$

$$= [h'(D_1\Psi_1, D_2\Psi_2) + h'(D_2\Psi_1, D_1\Psi_2)$$

$$- h(\Psi_1, D\ \Psi_2)]dp \qquad (5.6)$$

Now,

$$h(D\ \Psi_1, \Psi_2)dp = [h(D_2^* D_1\Psi_1, \Psi_2)$$

$$+ h(D_1^* D_2\Psi_1, \Psi_2)]dp = [h(\Psi_2, D_2^* D_1\Psi_1)$$

$$+ h(\Psi_2, D_1^* D_2\Psi_1)]^* dp = [h'(D_2\Psi_2, D_1\Psi_1)$$

$$+ h'(D_1\Psi_2, D_2\Psi_1)]^* dp + d\theta_2(\Psi_2, D_1\Psi_1)$$

$$+ d\theta_1(\Psi_2, D_2\Psi_1) = [h(D_1\Psi_1, D_2\Psi_2)$$

$$+ h'(D_2\Psi_1, D_1\Psi_2)]dp + d\theta_2(\Psi_2, D_1\Psi_1)$$

$$+ d\theta_1(\Psi_2, D_2\Psi_1) \tag{5.7}$$

Combining (5.6) and (5.7) gives:

$$[h(D\ \Psi_1, \Psi_2) - h(\Psi_1, D\ \Psi_2)]dp$$

$$= d\theta_1(\Psi_1, D_2\Psi_2) + d\theta_2(\Psi_1, D_1\Psi_2)$$

$$+ d\theta_2(\Psi_2, D_1\Psi_1) + d\theta_1(\Psi_2, D_2\Psi_1)$$

Set:

$$\omega(\Psi_1, \Psi_2) = \theta_1(\Psi_1, D_2\Psi_2) + \theta_2(\Psi_1, D_1\Psi_2)$$

$$+ \theta_2(\Psi_2, D_1\Psi_1) + \theta_1(\Psi_2, D_2\Psi_1) \tag{5.8}$$

Then, we have:

$$d\omega(\Psi_1, \Psi_2) = [h(D\ \Psi_1, \Psi_2)$$

$$- h(\Psi_1, D\ \Psi_2)]dp \tag{5.9}$$

Relation (5.9) then shows how a "conserved current"

ω is constructed from the Lagrangian L.

6. CONSERVED CHARGE DENSITY CURRENTS CONSTRUCTED FOR THE DIRAC AND KLEIN-GORDON EQUATIONS

Suppose L is the simply connected Lorentz group, $\sigma(L)$ is a representation by L via operators on a complex vector space V, and $\Gamma(E)$ denotes the space of map ψ: $R^4 \to V$. Let D be the following linear differential operator: $\Gamma(E) \to \Gamma(E)$

$$D(\psi) = \gamma^{\mu\nu}\partial_\mu\partial_\nu\psi + \gamma^\mu\partial_\mu + \gamma \tag{6.1}$$

The $(\gamma^{\mu\nu}, \gamma^\mu, \gamma)$ are linear operators: $V \to V$. In order that D be invariant under the Lorentz group, the $(\gamma^{\mu\nu})$; (γ^μ); (γ) transform, under $\sigma(L)$, like "symmetric, second order tensors", like "vectors", and like "scalars".

In this section, we will present a "standard" way of writing down conserved "charge-density currents" for this equation. In turn, those currents enable one to make $\Gamma(E, D)$ into a Hilbert space, and to proceed along standard directions to the "second quantization" procedure, and then to the construction of "quantum field theory", at least for the "free field" case.

To start the ball rolling, we will suppose that the following further assumption holds:

There is a Hermitian symmetric form

$$(v_1, v_2) \to h(v_1/v_2) \text{ on } V \tag{6.2}$$

which is invariant under $\sigma(L)$, i.e.

$$h(\sigma(\ell)v_1/\sigma(\ell)v_2) = h(v_1/v_2)$$

for $v_1, v_2 \in V$.

If $A\colon V \to V$ is a linear transformation, A^* denotes the "Hermitian adjoint" of A with respect to the form (6.2):

$$\langle Av_1/v_2\rangle = \langle v_1/A^* v_2\rangle \tag{6.3}$$

for $v_1, v_2 \in V$

Now, for $\psi_1, \psi_2 \in \Gamma(E)$, set:

$$\omega_\mu(\psi_1, \psi_2) = h(\gamma^{\mu\nu}\partial_\nu\psi_1/\psi_2)$$
$$+ h(\psi_1/\gamma^{\mu\nu}\partial_\nu\psi_2) + h(\gamma^\mu\psi_1/\psi_2) \qquad (6.4)$$
$$+ h(\psi_1/\gamma^\mu\psi_2)$$

Exercise: If $\gamma = -\gamma^*$ i.e. γ is skew-Hermitian with respect to form (6.2), show that

a) ω_μ is conserved, i.e.

$$\partial_\mu\omega_\mu(\psi_1, \psi_2) = 0$$

for ψ_1, ψ_2 which satisfy:

$$D\psi_1 = D\psi_2 = 0$$

b) If an inner product $\langle\psi_1/\psi_2\rangle$ is defined for $\Gamma(E, D)$ as follows

$$\langle\psi_1/\psi_2\rangle = \int\omega_0(\psi_1, \psi_2)(0, \vec{x})d\vec{x} \qquad (6.5)$$

then this inner product is Hermitian-symmetric, and invariant under the action of the Poincaré group on the space $\Gamma(E,D)$.

Thus, the results of this exercise will provide us with the desirable properties of the currents for the purposes of quantum field theory. Our next goal is to investigate the relation between the possible positivity of the inner product (6.5), the possible positivity of the "energy", and the existence of a "charge conjugation" operator.

<u>DEFINITION</u>: With the above assumptions, consider $\Gamma(E, D)$, with the inner product defined by (6.5). The *energy* of a $\psi \in \Gamma(E, D)$ is the real number defined as follows:

$$\text{energy of } \psi = i\langle\partial_0\psi/\psi\rangle \tag{6.6}$$

We will now investigate the effects of the "discrete" symmetries on the "charge", $\langle\psi/\psi\rangle$, given by (6.5), and the "energy" of ψ, given by (6.6), for $\psi \in \Gamma(E, D)$. Let L_c denote again the "complexification" of L. Suppose that σ can be extended to give a representation of $\sigma(L_c)$ of L_c on V. (For example, this can be done on general Lie principles if V is finite dimensional.) Let $\ell_0 \in L_c$ be the element such that:

a) $\ell_0{}^2 = 1.$

b) ℓ_0 is in the center of L_c (6.7)

c) In terms of the homomorphism of $L_c \to SO(4, C)$, ℓ_0 goes into the Lorentz transformation: $x \to - x$

Then,

$$\begin{aligned}
\sigma(\ell_0)\gamma^{\mu}\sigma(\ell_0)^{-1} &= - \gamma^{\mu} \\
\sigma(\ell_0)\gamma^{\mu\nu}\sigma(\ell_0)^{-1} &= \gamma^{\mu\nu} \\
\sigma(\ell_0)\gamma\sigma(\ell_0)^{-1} &= \gamma
\end{aligned} \tag{6.8}$$

$\sigma(\ell)\sigma(\ell_0) = \sigma(\ell_0)\sigma(\ell)$ for all $\ell \in L$.

Let $\rho(\ell_0)$ be the following transformation: $\Gamma(E) \to \Gamma(E)$

$$\rho(\ell_0)(\psi)(x) = \sigma(\ell_0)\psi(-x)$$

for $\psi \in \Gamma(E)$, $x \in R^4$

Then,

$$\rho(\ell_0)D = D\rho(\ell_0)$$

In particular, $\rho(\ell_0)$ maps $\Gamma(E, D)$ into itself. Let us now examine how it affects the inner product $< / >$ on $\Gamma(E, D)$.

<u>THEOREM 6.1</u>. Suppose that the Lie algebra of operators generated by $\sigma(L)$, the $\gamma^\mu + \gamma^{\mu *}$, and the $\gamma^{\mu\nu} + \gamma^{\mu\nu *}$ acts irreducibly on V.
Then

$$\sigma(\ell_0)\sigma(\ell_0)^* = \pm 1. \tag{6.9}$$

<u>Proof</u>. Since $\sigma(L)$ preserves the form h, and the adjoint operator * is taken with respect to h, we see that the operators $\gamma^\mu + \gamma^{\mu *}$ and $\gamma^{\mu\nu} + \gamma^{\mu\nu *}$ transform under $\sigma(L)$ like vectors and symmetric, second rank tensors. In particular, these relations are maintained when the real parameters of L become complex, hence these operators transform similarly under $\sigma(\ell_0)$. In particular,

$$\sigma(\ell_0)(\gamma^\mu + \gamma^{\mu *})\sigma(\ell_0) = - \gamma^\mu - \gamma^{\mu *}$$

$$\sigma(\ell_0)(\gamma^{\mu\nu} + \gamma^{\mu\nu *})\sigma(\ell_0) = \gamma^{\mu\nu} + \gamma^{\mu\nu *}$$

Taking the Hermitian adjoint of those relations gives

$$\sigma(\ell_0)^*(\gamma^\mu + \gamma^{\mu *})\sigma(\ell_0)^* = -\gamma^\mu - \gamma^{\mu\nu *}$$

$$\sigma(\ell_0)^*(\gamma^{\mu\nu} + \gamma^{\mu\nu *})\sigma(\ell_0)^* = \gamma^{\mu\nu} + \gamma^{\mu\nu *}$$

Similarly,

$$\sigma(\ell)\sigma(\ell_0)^* = \sigma(\ell_0)^*\sigma(\ell)$$

for all $\ell \in L$.

(Recall that $\sigma(\ell)^* = \sigma(\ell^{-1})$ for $\ell \in L$).

In particular, we see that

$$\sigma(\ell_0)\sigma(\ell_0)^*$$

commutes with $\sigma(L)$, $\gamma^\mu + \gamma^{\mu *}$ and $\gamma^{\mu\nu} + \gamma^{\mu\nu *}$. By Schur's lemma (since these operators act irreducibly),

$$\sigma(\ell_0)\sigma(\ell_0)^* = \lambda$$

where λ is a complex number.

Then, $\frac{1}{\lambda} = \sigma(\ell_0{}^{-1})^*\sigma(\ell_0{}^{-1}) = \sigma(\ell_0)^*\sigma(\ell)$

Then,

$$\lambda^2 = \sigma(\ell_0)\sigma(\ell_0)^*\sigma(\ell_0)\sigma(\ell_0)^*$$

$$= \frac{1}{\lambda}\,\sigma(\ell_0)\sigma(\ell_0)^* = 1.$$

This proves (6.9).

Thus, we analyze the possibilities as follows:

<u>Case 1</u>. $\sigma(\ell_0)\sigma(\ell_0)^* = 1.$

Then, $\sigma(\ell_0)$ preserves the form h. Hence, $\rho(\ell_0)$ preserves the form $<\psi/\psi>$ on $\Gamma(E, D)$. However, we have:

$$\rho(\ell_0)\delta_0\rho(\ell_0)^{-1} = -\,\delta_0 \tag{6.10}$$

Hence,

$$\text{energy of } \psi = -\text{ energy of } \rho(\ell_0)(\psi) \tag{6.11}$$

In particular, the energy cannot be positive.

<u>Case 2</u>. $\sigma(\ell_0)\sigma(\ell_0)^* = -\,1.$

Then

$$h(\sigma(\ell_0)v_1/\sigma(\ell_0)(\nu_2) = - h(v_1/v_2) \qquad (6.12)$$

for $v_1, v_2 \varepsilon V$.

Hence,

$$<\rho(\ell_0)\psi/\rho(\ell_0/\psi> = - <\psi/\psi> \qquad (6.13)$$

for $\psi \varepsilon \Gamma(E_1 D)$

The "charge density" cannot be positive. However, the "energy" might still be positive, since we have, combining (6.12) and (6.13)

energy of ψ = energy of $\rho(\ell_0)(\psi)$

In any case, we see that the "energy" and "charge" cannot be both positive (in case V is finite dimensional). The result was proved by Pauli. (See Gel'fand, Minlos and Shapiro [1] or Naimark [1] for a discussion.)

Exercise. Examine which of the alternatives holds

in the case where M is the Dirac or Klein-Gordon equations. Also, examine to the higher-spin wave equations. (See Umezawa [1].)

7. THE CHARGE-CONJUGATION OPERATION

Suppose that V, $\sigma(L)$, D, ω_μ are as in Section 6. So far we have only been concerned with the effect of the complex-linear transformations on V. In this section, we will consider anti-complex linear transformations on V, and their effect on $\Gamma(E, D)$, the space of solutions of the operation: $D\psi = 0$.

<u>DEFINITION</u>. A real-linear transformation

$A: V \to V$ is *anti-(complex) linear* if:

$$A(cv) = c^* v \tag{7.1}$$

for $v \in V$, $c \in C$.

Suppose that D is of the following form:

$$D = \gamma^\mu \partial_\mu + i\, m, \tag{7.2}$$

with m a real constant.

Then

$$DA = \gamma^\mu A\partial_\mu - A(im)$$

Then, the condition that A map $\Gamma(E, D)$ into itself is:

$$A^{-1}\gamma^\mu A = -\gamma^\mu$$

Let ℓ_0 be the element of L_c such that:

$$\sigma(\ell_0)\gamma^\mu\sigma(\ell_0) = -\gamma^\mu;\ \ell_0{}^2 = 1 \tag{7.3}$$

Then,

$$\sigma(\ell_0)A \quad \text{and} \quad A\sigma(\ell_0) \tag{7.4}$$

commute with γ^μ

Let us further suppose that

The (γ^μ) form an irreducible set of operators on V. (7.5)

For example, (7.5) will be true for the Dirac equation (exercise).

Thus, $A' = \sigma(\ell_0)A$ is an anti-linear transformation: $V \to V$ which commutes with an irreducible set of operators. By Schur's lemma,

$$A'^2 = \lambda$$

with a complex number.

Further, if A'' is any other anti-linear transformation which commutes with the γ^μ, then

$$A'A''$$

is linear, commutes with the γ^μ, hence by Schur's lemma is a multiple of the identity. Thus, A' (if it exists) is essentially determined by the γ^μ up to a scalar multiple.

To construct such an A explicitly, in an important special case split up V as $V^+ + V^-$, with:

$$V^+ = \{v \in V: \ \sigma(\ell_0)v = v\}$$

$$V^- = \{v \in V: \ \sigma(\ell_0)v = -v\}$$

Then, the following relations hold:

$$\gamma^{\mu}(V^{+}) \subset V^{-}, \ \gamma^{\mu}(V^{-}) \subset V^{+}$$

$$\sigma(L)(V^{+}) \subset V^{+}$$

$$\sigma(L)(V^{-}) \subset V^{-}$$

Denote by σ^{+} and σ^{-} the representation of L in V^{+} and V^{-}.

Suppose that the following conditions are satisfied.

σ^{+} and σ^{-} are equivalent via an anti-linear isomorphism α: $V^{+} \to V^{-}$, i.e. (7.6)
$\sigma^{-}(\ell)(\alpha(v^{+})) = \alpha(\sigma^{+}(\ell)v^{+})$ for $v^{+} \in V^{+}$

Set:

$$A(v^{+} + v^{-}) = \alpha(v^{+}) + \alpha^{-1}(v^{-}) \tag{7.7}$$

for $v^{+} \in V^{+}$, $v^{-} \in V^{-}$.

Then,

$$\begin{aligned}\sigma(\ell)A(v^{+} + v^{-}) &= \sigma^{-}(\ell)\alpha(v^{+}) + \sigma^{+}(\ell)\alpha^{-1}(v^{-})\\ &= \alpha(\sigma^{+}(\ell)v^{+}) + \alpha^{-1}(\sigma^{-}(\ell)(v^{-}))\\ &= A(\sigma^{+}(\ell)v^{+} + \sigma^{-}(\ell)v^{-})\\ &= A\sigma(\ell)(v^{+} + v^{-})\end{aligned} \tag{7.8}$$

Then, A intertwines the action of $\sigma(L)$, also

$$A^2 = \text{identity} \tag{7.9}$$

Let us examine the conditions that (7.2) be satisfied, with A defined by (7.7).

$$\begin{aligned}\gamma^\mu(v^-) &= \gamma^\mu AA(v^-) \\ &= A\gamma^\mu A(v^-) \\ &= -\alpha^{-1}\gamma^\mu\alpha(v^-)\end{aligned} \tag{7.10}$$

Thus, (7.10) determines how γ^μ acts on V^-, in terms of how it acts on V^+. Thus, the γ^μ can be chosen first as an intertwining map: $(\frac{1}{2}, \frac{1}{2}) \otimes V^+ \to V^-$, and then defined as mapping $V^- \to V^+$ via (7.10).

For example, for the Dirac equation,

$$V^+ = (\tfrac{1}{2}, 0); \; V^- = (0, \tfrac{1}{2})$$

$$(\tfrac{1}{2}, \tfrac{1}{2}) \otimes (\tfrac{1}{2}, 0) = (0, \tfrac{1}{2}) \otimes (1, \tfrac{1}{2})$$

Then there is a unique intertwining map

$$(\tfrac{1}{2}, \tfrac{1}{2}) \otimes (\tfrac{1}{2}, 0) \to (0, \tfrac{1}{2}).$$

The map $(\frac{1}{2}, \frac{1}{2}) \otimes (0, \frac{1}{2}) \to (\frac{1}{2}, 0)$ determining the action of γ^μ on V^- is defined via (7.10). The needed map α is provided by the following result:

Exercise. Let σ, σ' be irreducible representations, labelled (s_1, s_2), (s_1', s_2') in the spinorial notations, of $L = SL(2, C)$. Then, σ is "anti-equivalent" to σ', i.e. there is an intertwining anti-linear isomorphism between the vector spaces on which $\sigma(L)$ and $\sigma'(L)$ operate, if and only if

$$s_1' = s_2, \; s_2' = s_1.$$

Return to the general case, supposing again that α: $V^+ \to V^-$ is an L-intertwining, anti-linear isomorphism. Suppose that $(v_1^+, v_2^+) \to \beta(v_1^+, v_2^+)$ is a complex-bilinear form on V^+ which is invariant under the action of $\sigma^+(L)$. (Recall that we have proved already that such a form exists.)

Define:

$$h(v_1^+ + v_1^-/v_2^+ + v_2^-)$$
$$= \beta(\alpha^{-1}(v_1^-), v_2^+) + \beta(\alpha^{-1}(v_2^-), v_1^+)^*$$

for $v_1^+, v_2^+ \varepsilon V^+; v_1^-, v_2^- \varepsilon V^-$.

Then, h is a Hermitian-symmetric form on V that is invariant under the action of $\sigma(L)$ (exercise). Thus, h is the form needed to construct the "currents" ω_μ, hence also the inner product < | > on $\Gamma(E, D)$. Further,

$$h(A(v^+ + v^-)/(A(v^+ + v^-))$$

$$= h(\alpha(v^+) + \alpha^-(v^-)/\alpha(v^+) + \alpha^{-1}(v^-))$$

$$= \beta(v^+, \alpha^{-1}v^-) + (v^+, \alpha^{-1}v^-)^*$$

$$= h(v^+ + v^-/v^+ + v^-).$$

Thus

$$h(Av/Av) = h(v/v) \qquad (7.12)$$

for $v \varepsilon V$

Condition (7.12) - together, with the anti-linear condition - means that A is an "anti-unitary" operator with respect to the form h(/).

Now, we can describe the main physical property of this operator A.

<u>THEOREM 7.1</u>. Let A: $V \to V$ be the anti-unitary operator constructed above. Let D be given by (7.2), with $\Gamma(E, D)$ the space of all solutions of the equation: $D = 0$. Let C: $\Gamma(E, D) \to \Gamma(E, D)$ be defined as follows:

$$(C\psi)(x) = A(\psi(x))$$

Then, for $\psi \in \Gamma(E, D)$,

$$\text{charge } (C\psi) \text{ energy } (C\psi)$$
$$= \text{charge } \psi \text{ energy } \psi$$

The proof of this is left as an exercise (at least the reader should do the case of the Dirac and Klein-Gordon equations. We will return to the question of "charge conjugation" in more detail in Volume III.)

CHAPTER IX

FOCK SPACE AND FREE QUANTUM FIELDS

1. SYMMETRIC AND SKEW-SYMMETRIC TENSORS

Let H be a complex vector space. Let $T^r(H)$ denote the tensor product of H with itself r-number of times, $r \geq 0$:

$$T^r(H) = H \otimes H \otimes \ldots \otimes H.$$

Let T(H) be the direct sum of these vector spaces:

$$T(H) = H \oplus T^1(H) \oplus T^2(H) \oplus \ldots \tag{1.1}$$

One can form a bilinear product:

$T^r(H) \vee T^s(H) \to T^{r+s}(H)$, as follows:

$$(\Psi_1 \otimes \ldots \otimes \Psi_r)(\Psi_1{}' \otimes \ldots \otimes \Psi_s{}')$$

$$= \Psi_1 \otimes \ldots \otimes \Psi_r \otimes \Psi_1{}' \otimes \ldots \otimes \Psi_s{}' \qquad (1.2)$$

for $\Psi_1, \ldots, \Psi_r, \Psi_1{}', \ldots, \Psi_s{}' \in H$.

This product defines a bilinear product

$$: T(H) \times T(H) \to T(H),$$

which makes T(H) into an associative algebra; it is called the *tensor algebra* associated with H.

Now, consider the ideal generated[1] in T(H) by all elements of the form:

$$\Psi_1 \otimes \Psi_1 - \Psi_2 \otimes \Psi_1 \qquad (1.3)$$

Denote this ideal by $I_S(H)$. The quotient of the algebra T(H) by $I_S(H)$ is then an associative

---

[1]Recall that an *ideal* in T(H) is a linear subspace I such that $IT(H) \subset I$; $T(H)I \subset I$. The ideal *generated by a set of elements* is the smallest ideal containing these elements. The quotient of an algebra by an ideal inherits a quotient algebra structure.

algebra, denoted by $S(H)$, and called the *symmetric algebra* of H or the *algebra of symmetric tensors* over H. The product operation in $S(H)$ is denoted by: o :

Exercise: The product operation o in $S(H)$ is commutative, i.e.

$$s_1 \circ s_2 = s_2 \circ s_1 \quad \text{for} \quad s_1, s_2 \in S(H).$$

Since the elements of form (1.3) that generate $I_S(H)$ are of homogeneous degree, $S(H)$ inherits from $T(H)$ a degree structure; denote by $S^r(H)$ the quotient of $T^r(H)$ under the quotient map: $o \to I_S(H) \to T(H) \to S(H) \to o$: In particular, H may be identified with $T^1(H)$ and $S^1(H)$.

Exercise: H, as a subspace of $S(H)$, generates $S(H)$, i.e. every element of $S(H)$ may be written as a product of elements of H.

Exercise: If (Ψ_i), $1 \leq i \leq n$, are a set of linearly independent elements of H, then the elements

$$\left\{ \Psi_{i_1} \circ \cdots \circ \Psi_{i_r} \right\}$$

are linearly independent, for

$$1 \leq i_1 \leq i_2 \leq \cdots \leq i_r \leq n.$$

Exercise: Using the preceding two exercises, write down a basis of S(H) in terms of a basis for H, in case H is finite dimensional. Explain why the components of elements of S(H) with respect to this basis transform "like symmetric tensors".

Now, we turn to the study of the algebra of skew-symmetric tensors over H, denoted by: A(H): Let $I_A(H)$ be the ideal of T(H) generated by all elements of the form:

$$\Psi_1 \otimes \Psi_2 + \Psi_2 \otimes \Psi_1 \tag{1.4}$$

for $\Psi_1, \Psi_2 \in H$.

Let $A(H) = T(H)/I_A(H)$ be the quotient of T(H) by this ideal. The quotient algebra structure in A(H) is denoted by: $\wedge$: $A^r(H)$ denotes the image of $T^r(H)$ in this quotient.

<u>Exercise</u>: $A^1(H)$ may be identified with H.

<u>Exercise</u>: A(H) is the direct sum of the subspaces

$$H, A^2(H), A^3(H), \ldots$$

<u>Exercise</u>: $\Psi_1 \wedge \Psi_2 = (-1)^{rs} \Psi_2 \wedge \Psi_1$

for $\Psi_1 \in A^r(H)$, $\Psi_2 \in A^s(H)$.

<u>Exercise</u>: H generates A(H) as an algebra.

<u>Exercise</u>: If (Ψ_i), $1 \leq i \leq n$, are linearly independent elements of H, then

$$\Psi_{i_1} \wedge \cdots \wedge \Psi_{i_r}$$

are linearly independent elements of $A^r(H)$, for $1 \leq i_1 < i_2 < \ldots < i_r \leq n$.

<u>Exercise</u>: If H is finite dimensional, construct a basis for $A^r(H)$, using the previous two exercises.

Explain why the elements of $A^r(H)$ "transform like skew-symmetric tensors".

Exercise: Compute the dimension of $A^r(H)$.

Exercise: Compute the dimension of $A(H)$.

2. HERMITIAN-SYMMETRIC INNER PRODUCTS

Let H continue to be a complex vector space.

DEFINITION. A *Hermitian-symmetric inner product* for H is a real-bilinear map $(\Psi_1, \Psi_2) \to \langle\Psi_1/\Psi_2\rangle$ such that:

$$\langle\Psi_1/\Psi_2\rangle = \langle\Psi_2/\Psi_1\rangle^*$$

$$\langle c\Psi_1/\Psi_2\rangle = c^*\langle\Psi_1/\Psi_2\rangle = \langle\Psi_1/c^*\Psi_2\rangle \qquad (2.1)$$

for $c \in C$; $\Psi_1, \Psi_2 \in H$.

$c \to c^*$ denotes the complex-conjugate operation on complex numbers.

Now, our job is to extend such an inner product to $T(H)$, $S(H)$ and $A(H)$. First, extend it to $T^r(H)$, $r \geq 1$, as follows:

$$<\Psi_1 \otimes \ldots \otimes \Psi_r/\Psi_1' \otimes \ldots \otimes \Psi_r'>$$
$$= <\Psi/\Psi_1'> \ldots <\Psi_r/\Psi_r'> \tag{2.2}$$

for $\Psi_1, \ldots, \Psi_r, \Psi_1', \ldots, \Psi_r' \in H$.

Exercise: Show that (2.2) is indeed a Hermitian-symmetric inner product.

Extend to T(H) as follows:

$$<T^r(H)/T^s(H)> = 0$$
$$\text{if } r \neq s. \tag{2.3}$$

On to A(H), S(H) now: $< | >$ extends to S(H) as follows:

$$<\Psi_1 \circ \ldots \circ \Psi_r/\Psi_1' \circ \ldots \circ \Psi_r'>$$
$$= \frac{1}{r!} \Sigma <\Psi_1/\Psi_{i_1}'> \ldots <\Psi_r/\Psi_{i_r}'> \tag{2.4}$$

The sum on (2.4) is over all permutations $(1, \ldots, r) \to (i_1, \ldots, i_r)$ of the first r-integers.

Exercise: Show that (1.8) defines a genuine

Hermitian-symmetric inner product for $S^r(H)$.

Extend to $S(H)$ by requiring that:

$$\langle S^r(H)/S^s(H)\rangle = 0 \quad \text{for} \quad r \neq s. \tag{2.5}$$

Define $\langle \mid \rangle$ on $A^r(H)$ as follows:

$$\begin{aligned} &\langle \Psi_1 \wedge \dots \wedge \Psi_r / \Psi_1{}' \wedge \dots \wedge \Psi_r{}'\rangle \\ &= \frac{1}{r!} \Sigma \pm \langle \Psi_1/\Psi_{i_1}{}'\rangle \dots \langle \Psi_r/\Psi_{i_r}{}'\rangle. \end{aligned} \tag{2.6}$$

Again, the summation is over all permutations $(1,\dots, r) \to (i_1,\dots, i_r)$ of the r integers. The sign $\pm$ is the signature of the permutations.

Exercise: Show that (2.6) defines a Hermitian symmetric inner product for $A^r(H)$.

Exercise: Suppose that (Ψ_i), $1 \leq i, j \leq n$, is an *orthonormal basis* for H, i.e.

$$\langle \Psi_i/\Psi_j\rangle = \delta_{ij}.$$

Show that the following vectors form an orthonormal basis for $S(H)$ and $A(H)$, respectively:

$$\sqrt{r!}\ \Psi_{i_1} \circ \ldots \circ \Psi_{i_r},\ 1 \le i_1 < \ldots < i_r \le n$$
$$r = 1, 2, \ldots \qquad (2.7)$$

$$\sqrt{r!}\ \Psi_{i_1} \wedge \ldots \wedge \Psi_{i_r},\ 1 \le i_1 < \ldots < i_r \le n. \qquad (2.8)$$

To understand better why these definitions are made, proceed as follows: Let α: $T(H) \to T(H)$ and β: $T(H) \to T(H)$ be defined as follows:

$$\alpha(\Psi_1 \otimes \ldots \otimes \Psi_r) = \frac{1}{r!} \Sigma\ \Psi_{i_1} \otimes \ldots \otimes \Psi_{i_r} \qquad (2.9)$$

$$\beta(\Psi_1 \otimes \ldots \otimes \Psi_r) = \frac{1}{r!} \Sigma \pm \Psi_{i_1} \otimes \ldots \otimes \Psi_{i_r}$$

for $\Psi_1, \ldots, \Psi_r \in H$. (2.10)

Again, in (2.9)-(2.10), the sum is over all permutations $(1, \ldots, r) \to (i_1, \ldots, i_r)$ of the r integers. $\pm$ in (2.10) is the *signature* of the permutation, i.e. is + 1 if the permutation is a even, i.e. a product of an even number of transpositions, and is -1 if it is odd.

Notice the right hand sides of (2.9) and

(2.10) are "symmetric" and "skew-symmetric tensors", respectively. Also:

$$\text{The image of } \alpha(\Psi_1 \otimes \ldots \otimes \Psi_r) \text{ in } S(H) \text{ is } \Psi_1 \circ \ldots \circ \Psi_r \tag{2.11}$$

$$\text{The image of } \beta(\Psi_1 \otimes \ldots \otimes \Psi_r) \text{ in } A(H) \text{ is } \Psi_1 \wedge \ldots \wedge \Psi_r . \tag{2.12}$$

Exercise: T(H) is a direct sum of

$I_S(H)$ and $\alpha(T(H))$.

Exercise: T(H) is a direct sum of

$I_A(H)$ and $\beta(T(H))$.

Exercise:

$$\begin{aligned} \langle I_S(H)/\alpha(T(H))\rangle &= 0 \\ \langle I_A(H)/\beta(T(H))\rangle &= 0. \end{aligned} \tag{2.13}$$

Exercise: The quotient maps: $\alpha(T(H)) \to S(H)$,

$\beta(T(H)) \to A(H)$, are isomorphisms between the Hermitian symmetric inner products defined by the above formulas on these spaces.

For example, if H is a *Hilbert space* with respect to the inner product, i.e.

$$<\Psi/\Psi> > 0 \quad \text{for} \quad \Psi \in H,$$

then these results imply that $I_S(H)^-$, the "orthogonal complement" of $I_S(H)$ in the Hilbert space T(H), can be identified *as a Hilbert space*, with S(H). Similarly, $I_A(H)^-$ is identified with A(H). This explains the "naturalness" of these definitions.

3. CREATION AND ANNIHILATION OPERATORS

Let H be a vector space, which comes with a Hermitian symmetric inner product $(\Psi_1, \Psi_2) \to$ $<\Psi_1/\Psi_2>$. As we have seen in Section 2, this form can then be extended in a natural way to T(H), S(H) and A(H).

The algebras T(H), S(H), A(H) are extremely important in mathematics. They also play an important role in quantum field theory, as we shall see after completing our study of the formalism. To prepare for this, we must pass over from the "Schrödinger picture", which involves the vectors of S(H) or A(H), to the "Heisenberg picture", which involves operators.

For each Ψ in H, define the operator of creation by Ψ, denoted by A_Ψ^+, as follows:

For S(H), $A_\Psi^+(\Psi_1 \circ \dots \circ \Psi_r)$

$$= \sqrt{r+1}\ \Psi \circ \Psi_1 \circ \dots \circ \Psi_r . \tag{3.1}$$

For A(H), $A_\Psi^+(\Psi_1 \wedge \dots \wedge \Psi_r)$

$$= \sqrt{r+1}\ \Psi \wedge \Psi_1 \wedge \dots \wedge \Psi_r . \tag{3.2}$$

(The factor $\sqrt{r+1}$ is inserted for reasons that will be apparent soon.) Hence, A_Ψ^+ acts on S(H) by mapping $S^r(H)$ into $S^{r+1}(H)$, and acts in A(H) by mapping $A^r(H)$ into $A^{r+1}(H)$.

We must use the concept of *adjoint operator*, in the following general form: Suppose that H and H' are two complex vector spaces, with Hermitian symmetric inner products, denoted by: $\langle\Psi_1/\Psi_2\rangle$ and

$<\Psi_1'/\Psi_2'>$. Suppose that A: $H \to H'$ is a linear map. Then, the *adjoint map*, denoted by A^*, is the map: $H' \to H$ such that:

$$<A\Psi/\Psi'> = <\Psi/A^*\Psi'> \qquad (3.3)$$

for $\Psi \in H$, $\Psi' \in H'$.

Since we have equipped H, A(H), S(H) with such inner products, we are prepared to compute adjoints of various operators on the spaces.

Define the operator of annihilation by Ψ, denoted by A_Ψ^-, as the adjoint operator A_Ψ^+. A_Ψ^- must then be separately determined for S(H) and A(H) using the Hermitian symmetric product defined for them.

$$<\Psi_1' \circ \ldots \circ \Psi_{r-1}' | A_\Psi^-(\Psi_1 \circ \ldots \circ \Psi_1)>$$

$$<A_\Psi^+(\Psi_1' \circ \ldots \circ \Psi_{r-1}') | \Psi_1 \circ \ldots \circ \Psi_r>$$

$$= \sqrt{r} <\Psi \circ \Psi_1' \circ \ldots \circ \Psi_{r-1}' | \Psi_1 \circ \ldots \circ \Psi_r>$$

$$= \sqrt{r}\, \Sigma\, \frac{1}{r!} <\Psi | \Psi_{i_1}><\Psi_1' | \Psi_{i_2}> \ldots <\Psi_{r-1}' | \Psi_{i_r}>$$

Recall that the summation here is over all

permutations $(i_1, \ldots, r) \to (i_1, \ldots, r_r)$ of r integers. However, it can also be written as:

$$\sqrt{r}\,\frac{1}{r!}\sum_i \langle\Psi|\Psi_i\rangle \sum_{(i_2 \ldots i_r)} \langle\Psi_1{}'|\Psi_{i_2}\rangle \ldots \langle\Psi_{r-1}{}'|\Psi_{i_r}\rangle$$

The second summation is to be regarded as a permutation of the integers from 1 to r resulting from taking i out. Then, the second, "inside" summation, is just

$$(r-1)!\, \langle\Psi_1{}' \circ \ldots \circ \Psi_{r-1}{}'|\Psi_1 \circ \ldots$$

$$\Psi_{i-1} \circ \Psi_{i+1} \circ \ldots \Psi_r\rangle.$$

The whole summation is then:

$$\frac{1}{\sqrt{r}}\sum_i \langle\Psi|\Psi_i\rangle\langle\Psi_1{}' \circ \ldots \circ \Psi_{r-1}{}'|\Psi_i \circ \ldots$$

$$\Psi_{i-1} \circ \Psi_{i+1} \circ \ldots \circ \Psi_r\rangle.$$

Thus, we have:

$$A_\Psi{}^-(\Psi_1 \circ \ldots \circ \Psi_r) = \frac{1}{\sqrt{r}}\,(\langle\Psi|\Psi_1\rangle\Psi_2 \circ \ldots \circ \Psi_r$$

$$+ \langle\Psi|\Psi_2\rangle\Psi_1 \circ \Psi_3 \circ \ldots \circ \Psi_r$$

$$+\ldots+ \langle\Psi|\Psi_r\rangle\Psi_1 \circ\ldots\circ \Psi_{r-1}). \qquad (3.4)$$

A similar computation, left to the reader, shows that:

$$A_\Psi^-(\Psi_1 \wedge\ldots\wedge \Psi_r) = 1/\sqrt{r}\ (\langle\Psi|\Psi_1\rangle\Psi_2 \wedge\ldots\wedge \Psi_r$$

$$- \langle\Psi|\Psi_2\rangle\Psi_1 \wedge \Psi_3 \wedge\ldots\wedge \Psi_r +\ldots$$

$$+ (-1)^{r+1} \langle\Psi|\Psi_r\rangle\Psi_1 \wedge\ldots\wedge \Psi_r). \qquad (3.5)$$

For physical reasons that will be explained later, the operators A_Ψ^+ and A_Ψ^- acting in S(H) and A(H) will be called, respectively, *boson* and *fermion* creation and annihilation operators. Let us compute their commutation relations:

Boson case

$$A_\Psi^- A_{\Psi'}^+(\Psi_1\circ\ldots\circ \Psi_r) = \sqrt{r+1}\ A_\Psi^-(\Psi'\circ\Psi_1\circ\ldots\circ\Psi_r)$$

$$= (\langle\Psi|\Psi'\rangle\Psi_1\circ\ldots\circ\Psi_r + \langle\Psi|\Psi_1\rangle\Psi'\circ\Psi_2\circ\ldots\circ\Psi_r$$

$$+\ldots+ \langle\Psi|\Psi_r\rangle\Psi'\circ\Psi_1\circ\ldots\circ\Psi_{r-1}).$$

$$A_{\Psi'}{}^{+}A_{\Psi}{}^{-}(\Psi_1 \circ \ldots \circ \Psi_r)$$

$$= \frac{1}{\sqrt{r}} A_{\Psi'}{}^{+}(\langle\Psi|\Psi_1\rangle\Psi_2 \circ \ldots \circ \Psi_r + \ldots$$

$$+ \langle\Psi|\Psi_r\rangle\Psi_1 \circ \ldots \circ \Psi_{r-1})$$

$$= \langle\Psi|\Psi_1\rangle\Psi' \circ \Psi_2 \circ \ldots \circ \Psi_r + \ldots$$

$$\langle\Psi|\Psi_r\rangle\Psi' \circ \Psi_1 \circ \ldots \circ \Psi_{r-1}.$$

Thus

$$[A_{\Psi}{}^{-}, A_{\Psi'}{}^{+}](\Psi_1 \circ \ldots \circ \Psi_r)$$

$$= \langle\Psi|\Psi'\rangle\Psi_1 \circ \ldots \circ \Psi_r, \text{ or}$$

$$[A_{\Psi}{}^{-}, A_{\Psi'}{}^{+}] = \langle\Psi|\Psi'\rangle, \tag{3.6}$$

$$[A_{\Psi}{}^{+}, A_{\Psi'}{}^{+}] = 0 = [A_{\Psi}{}^{-}, A_{\Psi'}{}^{-}].$$

(The right hand side means the number $\langle\Psi|\Psi'\rangle$ times the identity operator, of course.)

Fermion case

$$A_{\Psi}{}^{-}A_{\Psi'}{}^{+}(\Psi_1 \wedge \ldots \wedge \Psi_r) = \sqrt{r+1}\, A_{\Psi}{}^{-}(\Psi' \wedge \Psi_1 \wedge \ldots \wedge \Psi_r)$$

$$= \langle\Psi|\Psi'\rangle\Psi_1 \wedge \ldots \wedge \Psi_r - \langle\Psi|\Psi_1\rangle\Psi' \wedge \Psi_2 \wedge \ldots \wedge \Psi_r + \ldots$$

$$A_{\Psi'}^{+}\ A_{\Psi}^{-}(\Psi_1\wedge\ldots\wedge\Psi_r)$$

$$= \frac{1}{\sqrt{r}}\ A_{\Psi'}^{+}(<\Psi|\Psi_1>\Psi_2\wedge\ldots\wedge\Psi_r$$

$$- <\Psi|\Psi_2>\Psi_1\wedge\Psi_3\wedge\ldots\wedge\Psi_r\ +\ldots)$$

$$= <\Psi|\Psi_1>\Psi'\wedge\Psi_2\wedge\ldots\wedge\Psi_r\ -\ <\Psi|\Psi_2>\Psi'\wedge\Psi_1\wedge\Psi_3\wedge\ldots\wedge\Psi_r+\ldots$$

Notice a peculiarity of the fermion case: $[A_\Psi^-, A_{\Psi'}^+]$ is nothing particularly simple or interesting, while $[A_\Psi^-, A_{\Psi'}^+]_+ = \frac{1}{2}\,(A_\Psi^- A_{\Psi'}^+ + A_{\Psi'}^+ A_\Psi^-)$ is simple, namely:

$$[A_\Psi^-, A_{\Psi'}^+]_+ = \frac{1}{2}\,<\Psi|\Psi'>. \tag{3.7}$$

Also, the annihilation and creation operators anticommute among themselves:

$$[A_\Psi^+, A_{\Psi'}^+]_+ = o = [A_\Psi^-, A_{\Psi'}^-]_+ \tag{3.8}$$

Notice the resemblance of the commutation relations (3.6) to the "Heisenberg" commutation relations. This can be made more explicit by constructing Hermitian operators from A_Ψ^+ and A_Ψ^-. Let

$$Q_\Psi = (A_\Psi{}^+ + A_\Psi{}^-)$$

$$P_\Psi = \frac{1}{i}(A_\Psi{}^+ - A_\Psi{}^-)$$

Then,

$$[Q_\Psi, Q_{\Psi'}] = [A_\Psi{}^+, A_{\Psi'}{}^-] + [A_\Psi{}^-, A_{\Psi'}{}^+]$$

$$= \langle\Psi|\Psi'\rangle - \langle\Psi'|\Psi\rangle \qquad (3.9)$$

$$[P_\Psi, P_{\Psi'}] = -(=[A_\Psi{}^+, A_{\Psi'}{}^-]$$

$$- [A_\Psi{}^-, A_{\Psi'}{}^+]) = -\langle\Psi'|\Psi\rangle + \langle\Psi|\Psi'\rangle \qquad (3.10)$$

$$[P_\Psi, Q_{\Psi'}] = \frac{1}{i}([A_\Psi{}^+, A_{\Psi'}{}^-]$$

$$- [A_\Psi{}^-, A_{\Psi'}{}^+]) = i(\langle\Psi'|\Psi\rangle + \langle\Psi|\Psi'\rangle) \qquad (3.11)$$

In particular, if $\langle\Psi|\Psi\rangle = 1$, notice that $P = \frac{P_\Psi}{\sqrt{2}}$, $Q = \frac{Q_\Psi}{\sqrt{2}}$ datisfy the Heisenberg commutation relations:

$$[P, Q] = i.$$

Similarly, if $\Psi_1, \ldots, \Psi_n$ is an orthonormal basis of H, then $P_j = P_{\Psi_j}/\sqrt{2}$, $Q_k = Q_{\Psi_k}/\sqrt{2}$ satisfy the n-dimensional version of the Heisenberg relations:

$$[P_j, P_k] = o = [Q_j, Q_k]$$
$$[P_j, Q_k] = i\delta_{jk},\ 1 \leq j,\ k \leq n. \tag{3.12}$$

(3.12) are called the "canonical" commutation relations. Thus the route is open to us to "quantize" systems with an infinite number of degrees of freedom by constructing these annihilation and creation operators for infinite dimensional Hilbert spaces. This is the great advantage of Hilbert space theory. The formalism (and many of the theorems) do not depend on finite dimensionality.

This specific realization of the "canonical" commutation relations is called the *Fock space representation*. If H is finite dimensional, any[1] realization of the canonical relations by Hermitian operators is equivalent to this one. However, in the infinite dimensional case - which is the case of interest for quantum field theory - this is not so.

---

[1]Modulo the usual fussy functional - analysis details, of course.

4. THE FOCK SPACE CONSTRUCTION OF FREE QUANTUM FIELDS

Let G be the "physical" Poincaré group, the semidirect product L·T of the simply connected Lorentz group L = SL(2, C), and a four dimensional translation group T·G then acts in its natural geometric way on R^4, where "R^4" is identified with the set of all real "space-time" vectors $x = (x_\mu)$, $0 \leq \mu \leq 3$, with: x_0 = time:, :x_i:, $1 \leq i, j \leq 3$, the "space" components.

Let $E \to R^4$ be a G-homogeneous vector bundle over "space-time". Then, as we have seen, Γ(E), the space of its cross-sections, can be identified with the space of all maps Ψ: $R^4 \to V$, where V is a vector space (the fiber of E over the point "zero" in R^4). The action of L on the vector bundle determines a representation σ(L) by linear operator on V. This determines an action ρ(L) of L on Γ(E) as follows:

$$\rho(\ell)(\Psi)(x) = \sigma(\ell)\Psi(\ell^{-1}x). \qquad (4.1)$$

Let D: $\Gamma(E) \to \Gamma(E)$ be a constant coefficient (i.e., translation-invariant) linear differential operator, with $\Gamma(E, D)$ denoting the space of $\Psi \in \Gamma(E)$ such that:

$$D\Psi = 0.$$

Let $\Gamma_0(E, D)$ and $\Gamma_0(E)$ denote the subspaces consisting of the cross-sections with compact support.

We must now describe how to construct a Hermitian-symmetric inner product on $\Gamma_0(E, D)$. We will now slightly modify the situation described in Chapter VIII.

Suppose $(v_1, v_2) \to h(v_1/v_2)$ is a Hermitian symmetric inner product on V that is invariant under the action of $\sigma(L)$, i.e.

$$h(\sigma(\ell)(v_1)/\sigma(\ell)(v_2)) = h(v_1/v_2) \qquad (4.2)$$

for $v_1, v_2 \in V$.

Suppose that D_μ: $\Gamma(E) \to \Gamma(E)$ are a set of four linear differential operators. Let

$$\omega_\mu(\Psi_1, \Psi_2) = h(D_\mu\Psi_1, \Psi_2) + h(\Psi_1, D_\mu\Psi_2) \qquad (4.3)$$

for $\Psi_1, \Psi_2 \in \Gamma(E)$.

Notice that:

$$\omega_\mu(\Psi_1, \Psi_2)^* = \omega_\mu(\Psi_2, \Psi_1). \tag{4.4}$$

Suppose now that:

The operators "D_μ" transform "like a four-vector" under the action of $\rho(L)$ on $\Gamma(E)$. (4.5)

$$\partial_\mu \omega_\mu(\Psi_1, \Psi_2) = 0 \tag{4.6}$$

for $\Psi_1, \Psi_2 \in \Gamma(E, D)$.

Conditions (4.5)-(4.6) means that "ω_μ is a conserved, Lorentz-invariant current".

The (ω_μ) also can be regarded from the more general, differential-geometric point of view described in Chapter VIII. For $\Psi_1, \Psi_2 \in \Gamma(E)$, set:

$$\begin{aligned} \omega(\Psi_1, \Psi_2) &= \omega_0(\Psi_1, \Psi_2)dx_1 \wedge dx_2 \wedge dx_3 \\ &- \omega_1(\Psi_1, \Psi_2)dx_0 \wedge dx_2 \wedge dx_3 \\ &+ \omega_2(\Psi_1, \Psi_2)dx_0 \wedge dx_1 \wedge dx_3 \\ &- \omega_3(\Psi_1, \Psi_2)dx_0 \wedge dx_1 \wedge dx_2. \end{aligned} \tag{4.7}$$

Thus, ω is a bilinear differential operator

$$\Gamma(E) \times \Gamma(E) \to F_c^3(R^4).$$

Conditions (4.2)-(4.5) then guarantee that this map intertwine the action of the Poincaré group on those cross-section-of-vector-bundle spaces. Condition (4.6) means that:

$$d\omega(\Psi_1, \Psi_2) = 0 \quad \text{if} \quad \Psi_1, \Psi_2 \in \Gamma(E, D). \tag{4.8}$$

In turn, (4.8) implies that an "inner product" for Γ(E, D) can be defined in the following way:

Choose a 3-dimensional submanifold N of R^4, and set:

$$<\Psi_1/\Psi_2> = \int_N \omega(\Psi_1, \Psi_2). \tag{4.9}$$

Then, the "conserved" condition (4.8) - together with Stokes' theorem, assuming of course that it can be applied - guarantees that (4.9) is actually independent of the submanifold N chosen, so long as it varies within a class that "cobound".

In particular, we can choose N as the submanifold: $x_0 = 0$. Then, (4.9) takes the following form:

$$\langle\Psi_1/\Psi_2\rangle = \int \omega_0(\Psi_1, \Psi_2)(0, \vec{x})d\vec{x}. \qquad (4.10)$$

An alternate notation for the right hand side of (4.10) might be:

$$\int \omega_0(\Psi_1, \Psi_2)/_{x_0=0}\, dx.$$

Of course if N is a submanifold of the form:

$$N = \{x \in R^4 \; x\cdot y = c\}$$

for some $y \in R^4$, $c \in R$, and if $y = (y_\mu)$, then, formally at least,

$$\langle\Psi_1/\Psi_2\rangle = \int g_{\mu\nu}y_\mu\omega_\nu(\Psi_1, \Psi_2)(x)/_{x\cdot y=c}\, dx \qquad (4.11)$$

where "dx" denotes the Euclidean-invariant volume element on the submanifold N of R^4.

At any rate, formulas of type (4.9)-(4.11) define a skew-Hermitian inner product $\langle \mid \rangle$ on

$\Gamma_0(E, D)$. G, the Poincaré group, then acts as a group of automorphisms of $\Gamma_0(E, D)$ with this form.

Suppose then we set: $H = \Gamma_0(E, D)$:, and construct the coresponding creation and annihilation operators, A_Ψ^+, A_Ψ^-, acting in the corresponding Fock space. (S(H) for bosons, A(H) for fermions). For $\Psi \in \Gamma_0(E, D)$, set

$$\phi(\Psi) = \frac{1}{\sqrt{2i}} (A_\Psi^+ - A_\Psi^-). \tag{4.12}$$

Note that:

$$\phi(i\Psi) = \frac{1}{\sqrt{2}} (A_\Psi^+ + A_\Psi^-) \tag{4.13}$$

hence:

$$\begin{aligned} A_\Psi^- &= \frac{1}{\sqrt{2i}} (\phi(\Psi) + i\phi(i\Psi)^*) \\ A_\Psi^+ &= \frac{1}{\sqrt{2}} (\phi(\Psi) - i\phi(i\Psi)^*). \end{aligned} \tag{4.14}$$

Then, in the Boson case:

$$[\phi(\Psi), \phi(\Psi')] = \frac{1}{2} (\langle\Psi/\Psi'\rangle - \langle\Psi/\Psi'\rangle^*) \tag{4.15}$$

for Ψ, $\Psi' \varepsilon \Gamma_0(E, D)$.

In the Fermion case:

$$[\phi(\Psi), \phi(\Psi')]_+$$

$$= -\frac{1}{2}(-[A_\Psi{}^+, A_{\Psi'}{}^-] - [A_\Psi{}^-, A_{\Psi'}{}^+] \qquad (4.16)$$

$$= \frac{1}{2}(<\Psi/\Psi'> + <\Psi/\Psi'>^*)$$

for Ψ, $\Psi' \varepsilon \Gamma_0(E, D)$.

Note that the right hand sides of (4.15) and (4.16) are the $\sqrt{-1}$ times the imaginary and real parts of $<\Psi/\Psi'>$, respectively. There are very interesting algebraic relations underlying this material that we will go into later.

Let us write (4.15) and (4.16) in the more usual physicist's form, using the ideas of "generalized function theory". Let (v_a), $1 \leq a \leq n$, be a basis for V.

Our goal is to define "operator-valued functions of space-time points", $\phi_a(y)$, with $y \varepsilon R^4$. Intuitively, $(\phi_a(y))$ are to represent the "components" of the field representing a particle that is concentrated in a very small neighborhood of the point y of space-time.

Let us assume the simplest general case; D is a first order differential operator, with the "Cauchy problem" solvable. Separate the 4-vector y as follows:

$$y = (t_0, \vec{y});\ t_0 \in R,\ \vec{y} \in R.$$

Let $\Psi_{a,y}$ be the "solution" of: D = 0 such that:

$$\Psi_{a,y}(t, \vec{x})/_{t=t_0} = v_a \delta(\vec{x} - \vec{y}). \tag{4.17}$$

Set:

$$\phi_a(y) = \phi(\Psi_{a,y}). \tag{4.18}$$

Suppose also that:

$$h(v_a/v_b) = h_{ab}. \tag{4.19}$$

$$D_\mu(\Psi(x)) = \beta^\mu(\Psi(x))/_{t=t_0}, \tag{4.20}$$

where β^μ are linear maps: $V \to V$, i.e. the D_μ are "zeroth order" differential operators.

Then, $y \to \phi_a(y)$ are in some generalized sense

"operator-valued functions of points of space time".

This is the form in which "free quantum fields" are presented in quantum field theory textbooks. Of course, one virtue of this way of looking at them is that one can attempt to treat "interacting fields" in the same fashion, i.e. as "operator-valued functions of points of space-time". If one, in addition, wants "fields" which have positive energy, one must take $\Gamma(E, D)$ to be the positive energy solutions of $:D\psi = 0:$

5. ANNIHILATION AND CREATION OPERATORS AND THE HEISENBERG LIE ALGEBRA

The "Fock space" constructions given in preceding sections have many interesting ramifications in mathematics and physics. In the remaining sections of this Chapter, we will present various remarks in this direction.

First, let us define the "Heisenberg algebra" algebraically. Let V be a real vector space, with $\omega\colon V \times V \to R$ a real-bilinear form on V which is skew-symmetric, i.e.

$$\omega(v_1, v_2) = -\omega(v_2, v_1) \qquad (5.1)$$

for $v_1, v_2 \in V$.

Define a Lie algebra $\underset{\sim}{H}$ as follows: As a vector space, $\underset{\sim}{H}$ is a direct sum: $V \oplus R$: Denote the generator of the "R" part by "1". Then, an element of $\underset{\sim}{H}$ is of the form: $v \oplus c \cdot 1$:, with $v \in V$, $c \in R$. $\underset{\sim}{H}$ is made into a Lie algebra as follows:

$$[v_1\ v'] = \omega(v_1\ v')$$

$$[1, v] = 0$$

for $v, v' \in V$.

Then, the subspace generated by "1" is in the center of $\underset{\sim}{H}$, i.e. $[\underset{\sim}{H}, 1] = 0$. Further,

$$[\underset{\sim}{H}, \underset{\sim}{H}] \subset \text{center } \underset{\sim}{H}. \qquad (5.2)$$

(5.2) implies that the Lie algebra $\underset{\sim}{H}$ is "nilpotent". The center of $\underset{\sim}{H}$ consists of "1", plus the space of all vectors $v \in V$ such that

$$\omega(v, V) = 0 \qquad (5.3)$$

If ω is *non-degenerate*, i.e. if v satisfying (5.3) implies $v = 0$, then the center of $\underset{\sim}{H} = (1)$.

Suppose ω is non-degenerate. Then, the Fock space construction gives a method for constructing representations of $\underset{\sim}{H}$ by skew-Hermitian operators, which we will now describe. To apply this method, one must choose a linear transformation $J: V \rightarrow V$ such that:

$$J^2 = -1 \tag{5.4}$$

$$\omega(v_1, Jv_2) = \omega(v_2, Jv_1) \tag{5.5}$$

for $v_1, v_2 \in V$.

Using (5.4), one can make V into a *complex* vector space, in the following way:

$$(a + bi)(v) = av + bJ(v) \tag{5.6}$$

for $v \in V$; $a, b \in R$

Exercise: Show that condition (5.4) is just the condition needed to make V into a complex vector space.

Notice that if one starts off with V as a

complex vector space, then regards it as a real vector space by ignoring the multiplication by pure imaginary numbers, J can be defined as the operation of multiplication by $i = \sqrt{-1}$. Thus, "J" is precisely that element of structure that must be added to a real vector space to keep track of the fact that it may be obtained by restricting the field of scalars.

Let us now suppose that (5.5) is satisfied. We find:

$$\langle v_1 | v_2 \rangle = \omega(v_1, Jv_2) + i\omega(v_1, v_2) \qquad (5.7)$$

for $v_1, v_2 \in V$.

<u>THEOREM 5.1</u>. (5.7) defines $\langle v_1 | v_2 \rangle$ as a Hermitian symmetric form with respect to the complex structure for V defined by (5.6).

<u>Proof</u>. It is clear that (5.7) defines $\langle\ |\ \rangle$ as a real-bilinear map: $V \times V \to C$.

$$\langle v_2 | v_1 \rangle = \omega(v_2, Jv_1) + i\omega(v_2, v_1)$$

= , using (5.1) and (5.5),

$$\omega(v, Jv_2) - \omega(v_1, v_2)$$

$$= \langle v_1 | v_2 \rangle^*$$

$$\langle iv_1 | v_2 \rangle = \langle Jv_1 | v_2 \rangle$$

$$= \omega(Jv_1, Jv_2) + i\omega(Jv_1, v_2)$$

$$= \omega(v_2, JJv_1) - i\omega(v_2, Jv_1)$$

$$= - \omega(v_2, v_1) - i\omega(v_1, Jv_2)$$

$$= \omega(v_1, v_2) - i\omega(v_1, Jv_2)$$

$$= - i\langle v_1 | v_2 \rangle$$

$$\langle v_1 | iv_2 \rangle = \langle v_1 | Jv_2 \rangle$$

$$= \omega(v_1, J^2 v_2) + i\omega(v_1, Jv_2)$$

$$= - \omega(v_1, v_2) + i\omega(v_1, Jv_2)$$

$$= i\langle v_1 | v_2 \rangle$$

These calculations show that $\langle \; | \; \rangle$ is Hermitian-symmetric.

Thus, we may adapt (5.7) to make V into a complex vector space with a Hermitian-symmetric form, and thereby construct the Boson annihilation and creation operators. Let $\phi(v)$, for $v \in V$, be the Hermitian operator defined by (4.12). Then, by (4.15),

$$[\phi(v_1), \phi(v_2)] = i\omega(v_1, v_2) \tag{5.8}$$

for $v_1, v_2 \in V$.

Thus, if we define:

$$\begin{aligned} &\rho(v) = i\phi(v), \\ &\rho(1) = -\,i \text{ times the (identity operator)} \end{aligned} \tag{5.9}$$

this assigns a skew-Hermitian operator (in S(V)) to each element of $\underset{\sim}{H}$. (5.8) implies that:

$$\begin{aligned} &[\rho(1), \rho(v)] = 0 \\ &[\rho(v_1), \rho(v_2)] = -\,i\omega(v_1, v_2) \\ &= \rho([v_1, v_2]) \end{aligned} \tag{5.10}$$

(5.10) then asserts that ρ defines a representation of $\underset{\sim}{H}$ by skew-Hermitian operators.

Let G be the group of linear isomorphisms $g: V \to V$ which preserve the form $\omega(\ ,\ )$, i.e.

$$\omega(gv_1, gv_2) = \omega(v_1, v_2)$$

for all $v_1, v_2 \in V$.

(If V is a vector space of dimension 2n, then G is denoted by Sp(n, R)). Then, g determines an automorphism of $\underset{\sim}{H}$, i.e. g applied to "1" is defined as "1". Thus, one can define a representation ρ_g of $\underset{\sim}{H}$ as follows:

$$\rho_g(v) = \rho(gv)$$

for $v \in H$.

In case V is finite dimensional, Stone, von Neumann and Weyl have proved that ρ_g is equivalent to ρ, i.e. there is a unitary operator A(g) such that:

$$\rho_g(v) = A(g)\rho(v)A(g)^{-1} \tag{5.11}$$

for all $v \in \underset{\sim}{H}$.

(Strictly speaking, this has been done only in case the Hermitian symmetric form $< \mid >$ is positive definite. A(g) is then an operator on the "completion" of the Hilbert space structure defined on S(V). D. Shale [1] has extended this to the case of an infinite dimensional V. It would be interesting to treat the case of a non-positive form $< \mid >$ as well.)

Thus, if $g_1, g_2 \in G$,

$$\rho_{g_1 g_2} = A(g_1 g_2)\rho A(g_1 g_2)^{-1}$$

$$= A(g_1)A(g_2)\rho A(g_2)^{-1}A(g_1)^{-1}.$$

Then, the operators $A(g_1 g_2)A(g_2)^{-1}A(g_1)^{-1}$ commute with the operators $\rho(\underset{\sim}{H})$. However, the operators $\rho(\underset{\sim}{H})$ are in some sense "irreducible". Hence one expects - via some infinite dimensional version of Schur's lemma - that $A(g_1 g_2)A(g_2)^{-1}$ $A(g_1)^{-1}$ is a scalar multiple of the identity. Thus, there should be some relation of the form:

$$A(g_1 g_2) = A(g_1)A(g_2)e^{i\lambda(g_1,\, g_2)} \tag{5.12}$$

with $\lambda(g_1, g_2)$ a real number. Thus, $g \to A(g)$ is a representation "up to a factor" or a "ray representation" of G.

In fact, Shale has shown that - at least in the positive definite case - $\lambda(g_1, g_2)$ can be chosen to be ± 1. Thus, $g \to A(g)$ is a "true" representation of a two-fold covering group of $Sp(n, R) = G$. Itzykson [1] has identified this two-fold covering group and discussed its representations.

We will not go into this very interesting material in any detail at this point. Instead, we will make several auxiliary remarks.

First, suppose that g: $V \to V$ is a *unitary* operator with respect to the Hermitian form $< | >$. The conditions for this are:

$$gJ(v) = J(gv) \tag{5.13}$$

for $v \in V$,

i.e. g is a complex-linear transformation: $V \to V$, and

$$\omega(gv_1, gv_2) = \omega(v_1, v_2) \tag{5.14}$$

for $v_1, v_2 \in V$.

i.e. $g \in G$. (5.13) and (5.14) together then imply

that

$$\langle gv_1 | gv_2 \rangle = \langle v_1 | v_2 \rangle \tag{5.15}$$

i.e. g is "unitary".

Since g preserves the complex structure, it extends naturally to a linear transformation: $T(V) \to T(V)$ of the tensor algebra, and preserves the ideals whose quotient is $S(V)$, hence g induces a complex-linear transformation $s(g)$: $S(V) \to S(V)$. This linear transformation also obviously preserves the Hermitian-symmetric form on $S(V)$. Further,

$$\begin{aligned} s(g)A_v^{+}\, s(g^{-1}) &= A_{gv}^{+} \\ s(g)A_v^{-}\, s(g^{-1}) &= A_{gv}^{-} \end{aligned} \tag{5.16}$$

for $v \in V$, with A_v^{+}, A_v^{-}: $S(V) \to S(V)$

the creation and annihilated operators.

Exercise. Prove (5.16)

Hence, we have:

$$s(g)\rho(v)s(g^{-1}) = \rho(gv) \tag{5.17}$$

for $v \in V$.

If, G_u denotes the subgroup of G consisting of the unitary transformations, then $g \to s(g)$ defines a representation of G_u by unitary operators on S(V) which is the restriction of the representation - up-to a - factor, (5.13), defined above.

A second type of elements of G are the *anti-unitary operators*. A $g \in G$ is called by this name if:

g, as a map $V \to V$, is anti-complex linear, i.e.

$$g(Jv) = - Jg(v) \tag{5.18}$$

for $v \in V$.

If $g \in G$ satisfies (5.18), let us compute:

$\langle gv_1 | gv_2 \rangle$ = , using (5.7),

$\omega(gv_1, Jgv_2) + i\omega(gv_1, gv_2)$

= using (5.18),

$- \omega(gv_1, gJv_2) + i\omega(v_1, v_2)$

$$= -\omega(v_1, Jv_2) + i\omega(v_1, v_2)$$

$$= -\langle v_1 | v_2 \rangle^* \tag{5.19}$$

Let us pause to consider the physical significance of this relation. A complex-linear vector space V, with a Hermitian symmetric form $\langle v_1 | v_2 \rangle$ may be considered as a system of "states" for a quantum mechanical system. Thus, if

$$\langle v_1 | v_2 \rangle = re^{i\theta}$$

with r, θ real numbers, then r represents some idea of "probability" for a "transition" between v_1 and v_2, while θ is some sort of "phase". (Of course, if the form is not positive definite, this is not quite standard - but perhaps such a generalization can be made). Then, a real-linear transformation A: $V \to V$ such that

$$|\langle Av_1 | Av_2 \rangle| = |\langle v_1 | v_2 \rangle| \tag{5.20}$$

$$\text{for} \quad v_1, v_2 \in V$$

preserves "probabilities", hence can play the role

of a physical "symmetry". This general mathematical picture of what a "symmetry" should mean has been emphasized by Wigner [1].

Returning to the main situation discussed in this section, we see that g satisfying (5.19) satisfies (5.20), hence falling under the broad category of possible "symmetries".

We can also consider the Lie algebra of $\underset{\sim}{G}$. It consists of the real linear transformations $X: V \to V$ such that:

$$\omega(Xv_1, v_2) + \omega(v_1, Xv_2) = 0 \tag{5.21}$$

for $v_1, v_2 \in V$.

For an $X \in \underset{\sim}{G}$, we can then assign a real-linear form β_X on V as follows:

$$\beta_X(v_1, v_2) = \omega(Xv_1, v_2) \tag{5.22}$$

Notice that (5.22), combined with the skew-symmetry of ω, implies that β_X is symmetric, i.e.

$$\beta_X(v_1, v_2) = \beta_X(v_2, v_1) \tag{5.23}$$

for $v_1, v_2 \in V$

In turn, (5.23) enables us to identify β_X with an element of $S(V^d)$, the symmetric algebra of the real vector space V^d (= the real dual of V). We will now pause to develop this point.

Suppose for the moment that V is merely a real vector space V^d, the *dual space to V* is defined as the vector space of all real-linear maps $\theta\colon V \to R$. V^d is a vector space over the real numbers. $T(V^d)$ denotes the *real* tensor algebra, while $S(V^d)$ denotes the algebra of symmetric tensors; recall that it is the quotient of $T(V^d)$ by the ideal generated by all second degree tensors of the form:

$$\theta_1 \otimes \theta_2 - \theta_2 \otimes \theta_1$$

for $\theta_1, \theta_2 \in V^d$.

Let $T^r(V^d)$ denote the space of r-th degree tensors over V^d. It can be identified with the vector space of all, real, r-multilinear maps

$$\beta\colon (v_1,\ldots, v_r) \to \beta(v_1,\ldots, v_r) \qquad (5.24)$$

of $V \times \ldots \times V \rightarrow R$. To see this, assign to $\theta_1 \otimes \ldots \otimes \theta_r$, with $\theta_1, \ldots, \theta_r \in V^d$, the following map of form (5.2):

$$\beta(v_1, \ldots, v_r) = \theta_1(v_1)\theta_2(v_2)\ldots\theta_r(v_r) \quad (5.25)$$

<u>Exercise</u>. Carry out the details of this identification of $T(V^d)$ with multi-linear maps on V.

Consider a map of form (5.24) that is *symmetric* under interchange of any of the arguments $v_1, \ldots, v_r$. Under the quotient homomorphism: $T(V^d) \rightarrow S(V^d)$, it has an image in $S(V^d)$. This determines a linear mapping of the symmetric maps of the form (5.24) into $S^r(V^d)$, i.e. the elements of $S(V^d)$ that are homogeneous of degree r.

<u>Exercise</u>. Show that this map of symmetric θ's of form (5.24) into $S^r(V^d)$ is an isomorphism. Compute how the "product" of two such symmetric θ's is to be computed so that this map is an algebra homomorphism. $(S(V^d))$ has a structure of associative algebra, since it is identified with the quotient of the tensor algebra $T(V^d)$ by an ideal. Recall that this product is commutative, and is denoted

by : o :

Now, suppose that, in addition, V has a skew-symmetric, nondegenerate bilinear form $(v_1, v_2) \to \omega(v_1, v_2)$. ω sets up an isomorphism of V with V^d in the following way:

Assign to $\theta \in V^d$ the vector $v(\theta) \in V$ such that:

$$\theta(v_1) = \omega(v(\theta), v_1) \tag{5.26}$$

for all $v_1 \in V$.

<u>THEOREM 5.2</u>. Given the form ω, $S(V^d)$ has a Lie algebra structure.

$$(\theta_1, \theta_2) \to \{\theta_1, \theta_2\}, \tag{5.27}$$

called *Poisson bracket*, such that:

$$\{a, \theta\} = 0 \tag{5.28}$$

for $a \in S^o(V^d) = R$, $\theta \in S(V^d)$

$$\{\theta_1, \theta_2\} = \omega(v(\theta_1)v(\theta_2)) \tag{5.29}$$

for $\theta_1, \theta_2 \in V^d$,

where $v(\theta_1)$, $v(\theta_2)$ are the elements of V defined by (5.26)

$$\{\theta_1, \theta_2 \circ \theta_3\} = \{\theta_1, \theta_2\} \circ \theta_3 + \theta_2 \circ \{\theta_1, \theta_3\} \tag{5.30}$$

$$\text{for} \quad \theta_1, \theta_2, \theta_3 \in S(V^d)$$

$$\{S^r(V^d), S^s(V^d)\} \subset S^{r+a-2}(V^d) \tag{5.31}$$

Exercise. Prove Theorem 5.2.

To justify the name "Poisson bracket", we must show how it reduces to this term as it is used in classical and quantum-mechanics. To see this, choose a basis $(v_i, v_i{}')$ of V, $1 \leq i, j \leq n$, such that:

$$\omega(v_i, v_j) = 0 = \omega_i(v_i{}', v_j{}') \tag{5.32}$$

$$\omega(v_i, v_j{}') = \delta_{ij}$$

Let (p_i, q_i) be the "dual basis" of V^d, i.e.

$$p_i(v_j) = 0 = q_i(v_j{}') \tag{5.33}$$

$$p_i(v_j{}') = \delta_{ij} = q_i(v_j)$$

Exercise. Show that ω - after identification with an element of $A^2(V^d)$ - is equal to

$$\pm\ p_i \wedge q_i \tag{5.34}$$

where r denotes the algebra product on $A(V^d)$ (= space of skew-symmetric tensors) inherited from $T(V^d)$ via the quotient. Determine which of the signs $\pm$ holds.

However, identified the p's and q's with elements of V^d, we can identify polynomials of (p, q) of degree r in the p's and q's with elements of $S^r(V^d)$.

Exercise. Show that the "Poisson bracket" defined in $S(V^d)$ by Theorem 5.2 agrees with the "classical" Poisson bracket,

$$\{f,\ f'\} = \frac{\partial f}{\partial p_i}\frac{\partial f'}{\partial q_i} - \frac{\partial f}{\partial q_i}\frac{\partial f'}{\partial p_i} \tag{5.35}$$

Exercise. Consider the map $X \to \beta_X$ of $\underset{\sim}{G}$ into $S^2(V^d)$. (5.31) indicates that $S^2(V^d)$ is a Lie subalgebra of $S(V^d)$. Show that this map is a Lie algebra homomorphism.

Exercise. Let $\underset{\sim}{H}$ be the Heisenberg Lie algebra $V \oplus R$. Then, $\underset{\sim}{G}$ acts as a Lie algebra of derivations of $\underset{\sim}{H}$. Let $\underset{\sim}{G}'$ be the semi-direct sum Lie algebra defined by this action. Show that $\underset{\sim}{G}'$ is isomorphic to $S^0(V^d) + S^1(V^d) + S^2(V^d)$.

Now, let us suppose that an operator J: $V \to V$ exists, enabling us to construct a Hermitian symmetric form on V, annihilation and creation operators, and hence also the representation $v \to \rho(v)$ of $\underset{\sim}{H}$ by skew-Hermitian operators on $S(V)$. (Here, $S(V)$ denotes the symmetric algebra V as a complex vector space, defined by J). Let us attempt to extend this representation to $S^2(V^d)$. For $\theta_1, \theta_2 \in V^d$, define:

$$\rho(\theta_1 \circ \theta_2) = \rho(v(\theta_1))\rho(v(\theta_2)) + \lambda i\omega(v(\theta_1), v(\theta_2)) \tag{5.36}$$

where λ is a real number.

Exercise. Determine the value of λ (if any) for which (5.36) determines a representation of the Lie algebra $S^2(V^d)$ (hence of $\underset{\sim}{G}$) by skew-Hermitian operators on S(V). For which values of λ does it determine a representation of $S^2(V^d)$, not necessarily skew-Hermitian? Show that (5.37) also determines a representation of the semi-direct sum algebra $\underset{\sim}{G} + \underset{\sim}{H}$. Show that ρ is the Lie-algebra representation corresponding to the representation - up-to - a factor representation of G discussed above.

6. CLIFFORD ALGEBRAS AND FERMIONS

In Section 5, we have sketched some mathematics associated with quantization-via-bosons. Now, we turn to the fermions case, where there are equally interesting ramifications.

Let V be a real vector space, and let $(v_1, v_2) \to \beta(v_1, v_2)$ be a *symmetric*, real-bilinear map: $V_1 \times V_2 \to R$. The *Clifford algebra* $C_\beta(V)$ (or C(V), if β is fixed throughout the discussion) is an associative algebra containing V as a subspace, and containing an "identity" element - denoted by "1" - such that:

$$v_1 v_2 + v_2 v_1 = 2\beta(v_1, v_2) \tag{6.1}$$

for $v_1, v_2 \in V$.

Explicitly, $C_\beta(V)$ is defined as follows:

Consider T(V), the associative algebra of real tensors over V. Let $I\beta(V)$ be the ideal generated by all elements of the following form:

$$v_1 \otimes v_2 + v_2 \otimes v_1 - 2\beta(v_1, v_2) \tag{6.2}$$

for $v_1, v_2 \in V$.

Notice that $\beta(v_1, v_2)$ is a real number. To consider it as an element of $T(V)_1$, we define $T^0(V)$ as the real numbers.

Then, $C_\beta(V)$ is defined as the quotient algebra $T(V)/C_\beta(V)$. The quotient product in $C_\beta(V)$ is denoted by v_1v_2. Notice that if β is identically zero, $C_\beta(V)$ reduces to A(V), the algebra of skew-symmetric tensors over V. However, if $\beta \neq 0$, the elements of form (6.2) do not have a homogeneous degree, hence $C_\beta(V)$ has no "grading" by degrees, as do the associative algebras (such as T(V), S(V), A(V)) we have considered up to now. However, V is embedded in $C_\beta(V)$ as a subspace.

Exercise. Show that $C_\beta(V)$ is isomorphis to A(V) as a *vector space*. Show that the dimension of $C_\beta(V)$ is 2^n, if V is n-dimensional.

Now, $C_\beta(V)$ is an associative algebra. Hence, we can construct a Lie algebra whose underlying vector space is $C_\beta(V)$, defining the Lie bracket as the commutator:

$$[v_1, v_2] = v_1 v_2 - v_2 v_1$$

$$\text{for} \quad v_1, v_2 \in C_\beta(V)$$

Let us compute a few commutation relations.

For $v_1, v_2 \in V$,

$$[v_1, v_2] = 2v_1 v_2 - 2\beta(v_1, v_2) \tag{6.3}$$

For $v_1, v_2, v \in V$,

$$[v_1 v_2, v] = v_1[v_2, v] + [v_1, v]v_2$$

$$= 2(v_1 v_2 v - v_1 \beta(v_2, v))$$

$$+ 2(v_1 v v_2 - \beta(v_1, v)v_2)$$

$$= 2v_1(- v v_2 + 2\beta(v_2, v)) - 2v_1\beta(v_2, v)$$

$$+ 2v_1 v v_2 - 2\beta(v_1, v)v_2$$

$$= 2v_1\beta(v_2, v) - 2v_2\beta(v_1, v) \tag{6.4}$$

Notice that (6.4) implies that $\mathrm{Ad}(v_1v_2)$ $:v \to [v_1v_2, v]:$ maps V into itself. Further, we have

$$\mathrm{Ad}(v_1v_2) = -\mathrm{Ad}(v_2v_1) \tag{6.5}$$

For $v, v' \varepsilon V$,

$$\beta(\mathrm{Ad}(v_1v_2)v, v') + \beta(v, \mathrm{Ad}(v_1v_2)(v'))$$

$$= 2\beta(v_1(v_2, v) - v_2\beta(v_1, v), v')$$

$$+ 2\beta(v_1\beta(v_2, v') - v_2\beta(v_1, v'), v)$$

$$= 2\beta(v_2, v)\beta(v_1, v') - 2\beta(v_1, v)(v_2, v')$$

$$+ 2\beta(v_1, v)\beta(v_2, v') - \beta(v_1, v')\beta(v_2, v)$$

$$= 0 \tag{6.6}$$

<u>Exercise</u>. Let $t \to \exp(t\mathrm{Ad}(v_1v_2))$ be the one parameter group of linear transformation of V generated by $\mathrm{Ad}(v_1v_2)$. Show that each transformation group

belongs to the group G of linear transformation of $g\colon V \to V$ such that:

$$\beta(gv_1, gv_2) = \beta(v_1, v_2),$$

that is, $\exp(t\mathrm{Ad}(v_1v_2)) \in G$ = groups of automorphisms of the form β.

<u>Exercise</u>: Show that the linear transformation of the form $\mathrm{Ad}(v_1v_2)$ span the Lie algebra of G.

Thus, we see one use of the Clifford algebra: It can be used to construct the Lie algebra of G in this way. Notice the analogy with the work of Section 5. There, we started off with a skew-symmetric form ω on V, and used it to identify V with V^d, and to construct a Lie algebra structure on $R \oplus V \oplus S^2(V)$ or, $R \oplus V^d \oplus S^2(V^d)$, given by "Poisson bracket". $S^2(V)$ was a subalgebra, and it acting on V realized the Lie algebra of the group of automorphisms of the form ω. Notice that (6.5) implies that $v_1 \wedge v_2 \to \mathrm{Ad}(v_1v_2)$ defines a linear mapping from $A^2(V)$ to $\underset{\sim}{G}$.

Let us now compute the commutation relations among the quadratic terms in $C_\beta(V)$. Suppose

$v_1, v_2, v_1'v_2' \in V$. Then

$$
\begin{aligned}
&[v_1v_2, v_1'v_2'] \\
&= [v_1v_2, v_1']v_2' + v_1'[v_1v_2, v_2'] \\
&= 2(v_1\beta(v_2, v_1') - v_2\beta(v_1, v_1'))v_2' \\
&+ 2v_1'(v_1\beta(v_2, v_2') - v_2\beta(v_1, v_2')) \\
&= 2(\beta(v_2, v_1')v_1v_2' - \beta(v_1, v_1')v_2v_2' \\
&+ \beta(v_2, v_2')v_1'v_1 - \beta(v_1, v_2')v_1'v_2 \qquad (6.7)
\end{aligned}
$$

This determines explicitly the Lie algebra structure among the quadratic terms in $C_\beta(V)$, i.e. the Lie algebra structure of $\underset{\sim}{G}$.

Now, we turn to the question of representing the associative algebra $C_\beta(V)$ by an algebra of operators on a complex vector space H. We will only deal with special sorts of V's, namely we suppose that:

a) $\beta(v_1, v_2)$ is a non-degenerate form
b) V is a complex vector space, i.e. there is a linear transformation $J: V \to V$ such that

$$J^2 = -1.$$

c) $\beta(v_1, v_2) = \omega(v_1, Jv_2)$,

where ω: $V \times V \to R$ is a non-degenerate, skew-symmetric, bilinear form on V.

(Of course, in view of b) a) can be turned around to define ω in terms of β.

$$\omega(v_1, v_2) = -\beta(v_1, Jv_2).$$

Then, as in Section 5, we can define an Hermitian symmetric form $< \mid >$ on V, as follows:

$$<v_1/v_2> = \beta(v_1, v_2) + i\omega(v_1, v_2).$$

Let H be A(V), the space of skew-symmetric *complex* tensors, on the *complex* vector space V. (Then, A(V) is the "fermion" Fock space.)

For $v \in V$, let $\phi(v)$ be the Hermitian operator on H constructed from the annihilation and creation operators. As we have seen:

$$[\phi(v_1), \phi(v_2)]_+ = 2\beta(v_1, v_2)$$

Thus, the map ϕ: $V + R \to$ (operators on H) can be extended to a map:

$$C_\beta(V) \to \text{(operators on H)} \tag{6.8}$$

Let us now explain how this map is constructed.

First, one can construct a representation of T(V) [1] by operators on H, as follows:

Assign to $v_1 \otimes \ldots \otimes v_r$ the operator

$$\phi(v_1 \otimes \ldots \otimes v_r) = \phi(v_1) \ldots \phi(v_r) \tag{6.9}$$

on H. Notice that (6.7) implies that every element of the ideal $I_\beta(V)$ defined by (6.2) goes into the zero operator (exercise). Thus, identifying $C_\beta(V)$ with $T(V)/I_\beta(V)$, we see that the map (6.8) can be defined as the quotient map associated with the map (6.9). Imposing the quotient associative algebra structure on $C_\beta(V)$, we see further that ϕ is a homomorphism of this algebra into the associative algebra of operators on the complex vector space H, i.e. ϕ is a linear representation of the

---

[1] Here, T(V) is the space of real tensors over the real vector space V.

Clifford algebra $C_\beta(V)$.

<u>Exercise</u>. If V is finite dimensional, show that $\phi(C_\beta(V))$ is an irreducible set of operators on H. If: dim V = 4:, and if β is a symmetric bilinear form whose canonical form has one plus, and three minus signs, identify $\phi(V)$ with the representation of the Clifford algebra given in physics books by the Dirac matrices. (Notice in this case that the dimension of V as a complex vector space is 2, hence the dimension of H as a complex vector space is $2^2 = 4$.)

Now, let G be the group of all *real* linear transformations g: $V \to V$ such that

$$\beta(gv_1, gv_2) = \beta(v_1, v_2)$$

for $v_1, v_2 \in V$

Again, define a presentation ϕ_g of V by operators on H as follows:

$$\phi_g(v) = \phi(g^{-1}v) \qquad (6.10)$$

for $v \in V$

As for the canonical commutation relation, one *expects* that ϕ_g and ϕ are unitarily equivalent, i.e. there is a unitary operator. $U(g)\colon H \to H$ such that

$$\phi_g(v) = \phi(g^{-1}v) = U(g)\phi(v)U(g)^{-1} \qquad (6.11)$$

for all $v \in V$.

Also, since $\phi(V)$ is irreducible, one expects that the assignment $g \to U(g)$ defines a representation of G "up to a factor", or a "true" representation of a covering group of G. This has been discussed as completely as is known by Shale and Stinespring [1]. If V is finite dimensional, the precise results have been known longer. Cartan [1] (see also Chevalley [1]). Then, there is a unique 2-fold simply connected covering group of G called the *spinor group* denoted by G'. The assignment $g \to U(g)$ defines a representation of G' called the *spinorial representation*. (It can be characterized group-theoretically as the representation in the smallest vector space which is a *faithful* representation of G', i.e. in which the center of G' does not go into the identity.) This representation was first discovered by E. Cartan in his classification of irreducibly representations of the real and complex Lie groups. After Dirac, it was recognized that it was definable in this very pretty way.

Again this assignment of *unitary* U(g) to a possibly-non-unitary g is of interest for the group-theoretic interpretation of the "discrete symmetries" of elementary particle physics. (For example, note that "charge conjugation is an anti-linear map on the single particle Hilbert space, but a unitary map on the full, field theory Hilbert space, where there are particles and anti-particles. (A relation analogous to (6.11) relates the two ideas.)

For example, suppose that $g\colon\ V \to V$ is anti-linear, and belongs to G. Then,

$$gJv = -\,Jgv$$

for $v \in V$.

Hence,

$$\omega(gv,\ gv_2) = -\,\beta(gv_1,\ Jgv_2)$$

$$= \beta(gv_1,\ gJv_2)$$

$$= \beta(v_1, Jv_2)$$

$$= -\omega(v_1, v_2) \qquad (6.12)$$

Then,

$$\langle gv_1/gv_2\rangle = \beta(v_1, v_2) - i\omega(v_1, v_2)$$

$$= \langle v_1/v_2\rangle^* \qquad (6.13)$$

Again, g as a map: $V \to V$: is "anti-unitary", in the sense used by Wigner, i.e. it preserves "probabilities" but not necessarily "phases". We hope to develop in a later volume these connections with the physics of elementary particles and their "discrete symmetries". This seems to be an area where sharpening the mathematical outlook might contribute new physical insights, particularly in view of the current difficulties with "CP violation".

BIBLIOGRAPHY

1. L. Auslander and R. Mac Kenzie, Introduction to Differentiable Manifolds, McGraw-Hill, New York, 1963.

1. M. Atiyah, K-Theory, W. A. Benjamin, New York, 1968.

1. V. Bargmann, On Unitary Ray Representations of Continuous Groups, Ann. of Math. 59, 1-46 (1954).

1. J. D. Bjorken and S. Drell, Relativistic Quantum Fields, McGraw-Hill, New York, 1965.

1. N. N. Bogoluibov and D. V. Shirkov, Introduction to the Theory of Quantized Fields, Interscience (New York), 1959.

1. P. A. M. Dirac, Lectures on Quantum Field Theory, Yeshiva University, 1966.

1. H. Flanders, Differential Forms, Academic Press, New York, 1963.

1. I. M. Gel'fand, R. A. Minlos, and Z. Y. Shapiro, Representations of the Rotation and Lorentz Group and Their Applications, Pergamon Press, New York, 1963.

1. K. Gottfried, Quantum Mechanics, W. A.

Benjamin, New York, 1966.

1. R. Hermann, Lie Groups for Physicists, W. A. Benjamin, New York, 1966.

2. _____, Differential Geometry and the Calculus of Variations, Academic Press, New York, 1968.

3. _____, Fourier Analysis on Groups and Partial Wave Analysis, W. A. Benjamin, New York, 1969.

4. _____, Lie Algebras and Quantum Mechanics, W. A. Benjamin, New York, to appear.

5. _____, Analytic Continuation of Group Representations, Comm. in Math. Phy.; Part 1, 2, 251-270 (1966); Part II, 3, 53-74 (1966); Part III, 3, 75-97 (1966); Part IV, Part V, 5, 157-190; (1967); Part VI, 6, 205-225, (1967).

6. _____, The Discrete Symmetries of Elementary Particle Physics, Ann. Inst. H. Poincaré Sect. A(N. S.), 7, 339-352 (1967).

1. L. Hörmander, Pseudo-Differential Operators, Comm. Pure Appl. Math. 18, 501-517 (1965).

1. C. Itzykson, Remarks on Boson Commutation Rules, Comm. Math. Phys. 4, 92-122, 1967.

1. N. Jacobson, Lie Algebras, Wiley (Interscience), New York, 1962.

1. R. Jost, The General Theory of Quantum Fields, American Math. Society, Providence, R. I., 1965.

1. T. D. Lee and G. C. Wick, Phys. Rev. 148, 1385 (1966).

1. L. Loomis and S. Sternberg, Advanced Calculus, Addison-Wesley, Reading, Mass., 1968.

1. G. Mackey, Induced Representations, W. A. Benjamin, New York, 1965.

1. M. Naimark, Linear Representations of the Lorentz Group, Macmillan, New York, 1964.

1. R. Palais, Seminar on the Atiyah-Singer Index Theorem, Princeton University Press, 1965.

1. ______, Global Analysis, W. A. Benjamin, New York, 1968.

1. J. Peetre, Une caracterisations abstraite des operateurs differentielles, Math. Scand. 7, 211-218, 1959.

1. W. Rühl, Complete Sets of Solutions of Linear Lorentz Covariant Field Equations With an Infinite Number of Field Components, Comm. Math. Phys., 6, 312-342.

1. S. S. Schweber, Relativistic Quantum Field Theory, Row, Peterson, Evanston, Ill., 1961.

1. D. Shale, Linear Symmetries of Free Boson Fields, Trans. American Math. Society, 103, 149-167, 1962.

1. D. Shale and W. F. Stinespring, States of the Clifford Algebra, Ann. of Math., 80, 365-381, 1964.

1. M. Spivak, Calculus on Manifolds, W. A. Benjamin, New York, 1965.

1. N. Steenrod, The Topology of Fiber Bundles, Princeton University Press, 1949.

1. R. F. Streater and A. S. Wightman, PCT, Spin and Statistics, and all that, W. A. Benjamin, New York, 1964.

1. M. Tinkham, Group Theory and Quantum Mechanics, McGraw-Hill, New York, 1966.

1. A. Trautman, Noether Equations and Conservation Laws, Comm. Math. Phys., 6, 248-261.

1. H. Umezewa, Quantum Field Theory, North-Holland, Amsterdam, 1956.

1. E. Wigner, Group Theory, Academic Press, New York, 1959.

2. _____, Unitary Representations of the Lorentz Group, Ann. Math., 40, 149, 1939.

1. C. N. Yang and R. N. Mills, The Conservation of Isotopispin and Isotopic Gauge Invariance, Phys. Rev. 96, 191-195 (1954).

1. B. Zumino and D. Zwanziger, Classification of Particle Multiplets, Phys. Rev. 1964, 1959-1981, (1967).